# ENERGY IN THE CITY ENVIRONMENT

Robert N. Rickles
Institute for Public Transportation
and
The New York Board of Trade, Inc.

NOYES PRESS
Park Ridge, New Jersey

Library of Congress Catalog Card Number: 72-96106
ISBN: 0-8155-5019-7

Published in the United States by
NOYES PRESS
Noyes Building
Park Ridge, New Jersey 07656

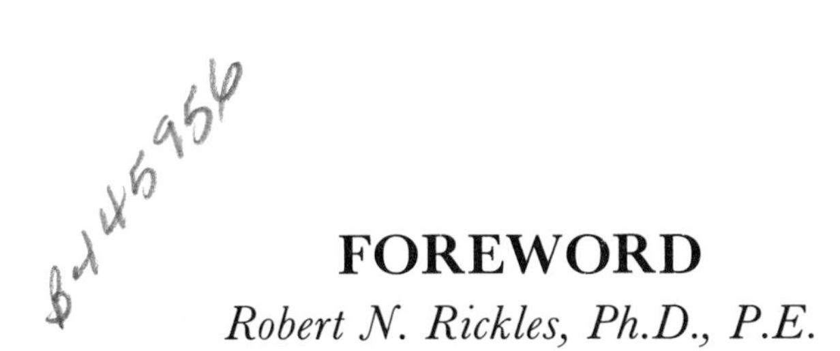

# FOREWORD

*Robert N. Rickles, Ph.D., P.E.*

Energy, the dictionary says, is the power by which anything acts effectively to move or change other things or accomplish any result . . . in science, the capacity of doing work and of overcoming inertia, as by heat, light, radiation, or mechanical and chemical forces.

Thus energy is necessary in undertaking any of the tasks of our society, large or small. All of our dreams and aspirations are bound up in the availability of energy, especially of low cost energy. And since the Industrial Revolution began, energy has always been readily and cheaply available on the American continent. No wonder it is so frightening to the American spirit and confidence to learn that we face rapidly increasing energy costs and a potential shortage of energy.

In one sense we have only ourselves to blame. For almost a century, we have wasted our extensive storehouse of energy without any thought that the gift is not boundless. With 6% of the world's population, we use almost 40% of the world's energy. If there is any area where energy is more extensively wasted than in the field of personal transportation, it would be surprising. We have built a system centered around the use of the automobile which uses 6 to 10 times more energy than comparable public transportation. The automobile and truck, in 1969, consumed some 90 billion gallons of gasoline and diesel fuel. Thus the simple movement of people and goods is a major consumer of energy resources. But so is virtually every segment of our national life. We are a society built on the availability and expenditure of cheap energy.

Suddenly we find that some real and some artificial shortages, the pressure brought to bear by environmentalists demanding cleaner fuels and restricting the means of producing them, foreign policy and defense considerations, real and imagined, together with the growing independence of oil producing countries and the growing dependence of the developed countries have all conspired to produce an air of crisis.

It is important that we recognize the dual level of this problem. There is first, the immediate problem that will be with us over the next two decades. Somehow, within the constrains of the current circumstances, we must provide the energy necessary for our society with the minimum environmental and economic dislocation.

But more to the point, we must recognize the long term problem. Energy, at least based on present technology, is not limitless and thus we must look to new ways of conducting ourselves in order to reduce the energy expenditures relative to our standard of living. We must, of course, recognize that many of these changes cannot make any significant impact upon energy demand for years if not decades, but if they are necessary we must begin now. We must also recognize that any such change in lifestyle must not prevent the necessary upward movement of millions of people who live in poverty in this country and of the billions around the world who live in peril of their existence daily.

Here in New York we have faced virtually every one of the energy problems of the country and in some cases to a greater degree than any other location. Our highly polluted and dangerous air has required the tightest of limitations on sulfur in fuel and the use of coal as an energy souce. Following the massive power blackout on the East Coast in 1965, we have seen blackouts and brownouts almost every summer with their attendant effects on the daily lives of our citizens.

We have lacked sufficient natural gas for expansion of its use and have been warned of cutbacks in this fuel and limitations on the availability of low sulfur fuel oil. We pay the highest prices in the country for energy and are afraid to permit the offshore search for oil for fear of another "Santa Barbara" while millions of our citizens go without jobs, heat and breath. We are in the middle of the energy crunch.

The hopeful note is that we continue to exist, continue to be the vibrant center of America. Perhaps even more hopeful is that leaders of industry, citizen groups and environmentalists could gather together under the auspices of the New York Board of Trade and the Council on the Environment to discuss and seek, in cooperation where possible, new solutions to our dilemma.

This book deals specifically with the energy problems of the New York Metropolitan Area, which includes one out of every ten Americans, but these problems are identical with those found in every urban area of every developed country. In that sense, the spokesmen for the various contending groups speak for their kind around the world, as they seek answers to the central problems. There are no right sides nor right answers. There are solutions, each one of which presents new and pressing problems. None of the final answers can properly come in a vacuum, none can come alone from one contending element or another. Each will require dialogue and discussion. That is the purpose of this volume—to discuss in an open fashion the problems, the potential answers and the drawbacks associated with them so that we can seek the path to a better, more humane world with as complete wisdom as is possible.

## Steering Committee

Alan L. Smith (Chairman)
Director of Public Relations
Brooklyn Union Gas Company
195 Montague Street
Brooklyn, New York 11201

Dorothy Job Donoghue
Director of Environmental Affairs
National Business Council on Environment
New York Board of Trade, Inc.
295 Fifth Avenue - Room 1218
New York, New York 10016

Mark Leymaster
Research Associate
Council on the Environment of New York City
51 Chambers Street
New York, New York 10007

Edward W. Livingston
Director, Community Relations
Consolidated Edison Company of N.Y.
4 Irving Place
New York, New York 10003

William Tell, Jr.
Associate General Counsel
Texaco Inc.
135 E. 42nd Street
New York, New York 10017

# CONTENTS

# WELCOME

*Neil H. Anderson*

On behalf of the National Business Council of the New York Board of Trade and the Council on Environment of New York City a most cordial welcome is extended to the representatives from government, business and the many civic and environmental organizations in New York.

The New York Board of Trade's Business Council on Environment was formed to further identify the cooperative roles of business, government and the public as related to the increasing environmental challenges facing not only this city but every city throughout the world. It has been our deep conviction that the time has come to direct our attention to the resources which will effect solutions to the already documented problems. We have repeatedly said that there are no solutions to problems within the problems. At best they only indicate a need for positive action and utilization of the technical and economic resources on these problems. We have successfully documented this premise and its orderly implementation. Leadership by example has been a by-word of the Board of Trade's activities both within the environmental areas and the broader social-economic projects in which business-government-public cooperation has been demonstrated.

Today we face an energy crisis—here and in every major city. Again, there is a need for understanding of not only the problems but, more important, what the resources are and how they can be better utilized. Through wise legislative and economic policies, public and environmental groups can have their rightful role in both determining priorities and insuring their full implementation.

We are indebted to the full cooperation and support of the

American Petroleum Institute, the Consolidated Edison Company and the Brooklyn Union Gas Company for their excellent guidance in the development of today's program which will focus on some of the immediate and longer range aspects of the energy crisis facing our city.

This is the first energy conference of this type to be held in our country. Today can be a milestone in demonstrating the effectiveness of cooperative participation between business, government and environmental representatives leading to both a meaningful dialogue and effective action.

The energy and environmental challenge is part of the universal search for quality in our physical, social and economic environments. We see an increasing awareness of the fact that our science and technology have given us unsurpassed quantities of goods and services. We must now be painfully but gratefully aware that the art of management and the appreciation of the qualities inherent in the individual hold the key to these qualities finding expression in the quantitative productive efforts both public and private.

## Introduction of Gordon C. Griswold

*Clifton Daniel*

We begin with Gordon C. Griswold, President of the Brooklyn Union Gas Company. He is prominent both in the gas industry and community affairs, which is a very good mix indeed for this occasion.

Nobody knows the Brooklyn Gas Company better than Mr. Griswold. He joined the company as a cadet engineer in 1933 and has been with Brooklyn Union ever since. He has long been a member of the American Gas Association, and has held high office in that and other gas industry organizations.

As for community affairs, Mr. Griswold is chairman of the Downtown Brooklyn Development Association, a member of the Brooklyn Chamber of Commerce, a director of the United Hospital Fund and chairman of its Brooklyn advisory committee, and Brooklyn chairman of the 1972 campaign of the United Fund of Greater New York.

As a citizen, in fact, he leads a double life—one life in Brooklyn, where he works; another in Westfield, New Jersey, where he lives. In Westfield he has also been active in the United Fund, once as President of the Board of Trustees, and as a member of the Westfield Town Planning Board.

As a good Brooklynite and good Jerseyite, he was obviously born somewhere else, namely in Denver, and was graduated from the Wharton School of Finance of the University of Pennsylvania.

Mr. Griswold—for the gas industry.

# ADDRESS

*Gordon C. Griswold*

## INTRODUCTION

We meet today to discuss two problems of major concern to New York and the nation as a whole. One deals with the supply of energy, and the other with the protection and improvement of the environment. Both are essential ingredients in the good life. Some of us are speaking to the energy problem and others to the environmental problem, but all of us are concerned with both. I, for example, will talk on behalf of the gas distribution of energy, but I am as interested as anyone in the room in breathing clean air, fishing in pollution-free waters, and preserving the natural beauty of the land. There seems to be a substantial opinion that the solutions to either problem must be at the expense of the other. This is not so. It is time to approach both the supply of energy and the protection of the environment objectively and constructively. Certainly, a meeting of this kind provides a good start in the direction of intelligent planning for and a practical resolution of these two vital problems so clearly in the true public interest.

The providing of energy for this nation, and particularly for this City, will be one of the most critical problems of the next 25 years. Equally as challenging will be the protection of the air quality, the marine life and the land as those energy supplies are made available.

And the necessary energy will be made available to New York City. There really is no other alternative other than serious decline. Commerce and Industry would move elsewhere leaving New York City without jobs and without hope. So the energy for growth is indeed an absolute necessity.

Meanwhile, every effort must be made to use every bit of energy intelligently. To this end we have made available a wide range of methods which result in energy savings.

## NATURAL GAS SHORTAGE

You have heard and read much about the "Energy Crunch." Others will tell you about oil and electricity, but it is no secret that natural gas is right now in short supply. For two years starting in 1969 our pipelines have been unable to provide additional supplies needed to supply the demand for clean fuel in New York City.

Last winter we suffered the first serious curtailment of pipeline contracts we had always considered to be firm obligations from the suppliers. Next winter those same pipelines will supply only 90% of the amount under contract. For natural gas, the energy shortage is here now.

This is particularly critical for New York City where natural gas appears to be the only available answer to our air pollution problems. Nationwide, over 50% of the natural gas sold is being used for heavy industrial uses or for electric generation simply because it is cheaper than other forms of fossil fuel. Yet here, where the quality of the air hangs in balance each day, we are unable to obtain—at any price—the volumes needed to protect the health and lives of the millions who live here.

There is adequate proof in the Federal and State air pollution hearings to document the dramatic part natural gas can and should play in the protection of the City environment. But what can be done to improve the supply?

## CONVENTIONAL SOURCES

There is stacked in Washington a mountain of paper testifying to the failure of the Federal Power Commission to adjust to changing conditions and provide flexible but fair regulation. In the 1950's and early 1960's, Brooklyn Union was one of the distributors who participated fully in the hearings which were aimed at setting fair and adequate area rates for natural gas.

Many mistakes were made in the ten years those hearings continued, but the most serious was the unbelievable delay in the FPC procedures in arriving at any decision at all. By the time some decisions finally came in the late 1960's, it was already too late. The indecision and uncertainty had caused the search for natural gas to decline because the funds available for exploration have been directed to the search for oil in foreign countries.

Figure 1. Specially Designed Tanker for Transporting Liquefied Natural Gas. (Courtesy of Brooklyn Union Gas Company)

Figure 2. Liquefied Natural Gas Terminal.
(Courtesy of Brooklyn Union Gas Company)

New approaches, painfully few and painfully slow though they are, do now show some promise for increases in the domestic supply. New approaches to price may now make it attractive for the producers to risk the huge capital necessary to do deeper drilling and develop marginal fields which can provide a much needed new horizon in domestic supply for the pipelines which supply New York City.

Our pipelines do not at this time reflect this optimism. They are predicting curtailment of existing contracts until 1976. This view makes it imperative that new sources of gas supply be rapidly explored.

## NEW SOURCES

One of the most promising unexplored areas in the world is that portion of the continental shelf which lies not far off the east coast of the United States. The lesson of the North Sea, where large quantities of natural gas were discovered so near to fuel-starved England, must not be lost on this country. The extreme importance of a fuel supply so close to the highly populated east coast demands that the existence of such a supply be proven or disproven. The very existence of large cities may hang in the balance and the future economy of the country may well hinge heavily on such a discovery. The East Coast continental shelf as a new potential source of energy simply must be explored.

There are large deposits of gas in Canada and in Alaska. The Canadians are understandably cautious about committing large volumes to export until they are certain about their own future energy needs. The Alaskan potential will begin to be realized when the pipeline considerations are finally settled. But New York City is unlikely to benefit directly because of the great distance from this source. We are keeping a wary eye on both these potentials.

## SUBSTITUTE SOURCES

Meanwhile, to assure the users of gas an adequate supply some unusual steps are being taken.

This winter, Brooklyn Union will receive 2.5 billion cubic feet of gas by barge from Everett, Massachusetts (see Figure 1). This will be liquefied natural gas from Algeria. The following year we will receive 4.5 billion cubic feet from a terminal now being constructed on Staten Island (see Figure 2). This dependence on foreign supplies is new to us but certainly not to the

Figure 3. Coal Gasification Pilot Plant.
(Courtesy of Institute of Gas Technology)

residual oil users of the east coast. Nonetheless, it is more uncertain and a more costly source than is the domestic supply. The need for these new sources is a clear indication of a new era in fuel and energy supply.

In 1973, Brooklyn Union expects to complete a Substitute Natural Gas Plant in Brooklyn. This plant, which reforms naphtha to produce methane, will be noiseless, odorless and pollution-free. It will produce 60 million cubic feet per day from naphtha which will be delivered directly to the plant by pipeline. The domestic source of this naphtha makes this a more secure energy source than would be true of imported naphtha.

Within the next ten years we look forward to plants which will produce methane from whole barrel crude oil and we are seriously considering a joint venture to develop such a plant.

Further down the time schedule is the gasification of coal. The Gas Industry and the Federal Government are cooperating in the lengthy process of experimentation and pilot plant work on this process (see Figure 3).

Nuclear stimulation of tight deposits, gas from oil shale or tar sands, or even the substitution of hydrogen are far-out possibilities.

These substitute sources will find their part in the overall energy picture providing better protection of supply under present uncertainties and good flexibility should the conventional supply picture improve.

## ENVIRONMENTAL CONSIDERATIONS

I would now like to spend a few minutes on natural gas as it relates to the total environment. The desirability of the methane molecule as a fuel is well proven. It burns completely and cleanly to carbon dioxide and water. Its ease of control makes possible lower level of nitrogen-oxides which occur to some extent when any fuel is burned with air. It is as close to an ideal environmental fuel as can be found with the possible exception of hydrogen which burns to only water vapor.

What then is the problem with natural gas drilling and why the delay in such vital areas as the offshore Louisana locations?

The only sin of natural gas is that it often occurs with oil. It is the fear of oil pollution from drilling platforms that causes delays and cancellations in essential lease sales. There has also been much effort, especially in New York State, to ban all drilling off the east coast.

We believe these efforts to be the product of misinformation.

The lawmakers and environmentalists who have been most concerned about such drilling are inadvertently proposing more foreign oil to fill the energy gap.

Data from the Department of Interior shows very clearly that by far the greatest danger of pollution is from tankers carrying oil and not from platforms. On this list, Santa Barbara, which is always pointed to as a disaster, ranks 31st in degree of severity. Tankers take all the pollution honors with the tanker Torrey Canyon, which foundered off the coast of England, ranking No. 1.

Platforms by definition are clearly more easily controlled. A daily fly-over by photo plane could clearly establish responsibility for any pollution. Tankers, most of which are foreign flag, and all of which are very transient, are not easily controlled. To repeat, by far the greatest danger is from tankers.

But even beyond all that, the people most adamant about drilling are guilty of the crime they are most vocal about. They do not consider the entire environment.

The full consideration of the environment must balance any possible pollution of offshore drilling against the benefits which natural gas can provide. We believe that the air which is breathed by millions of people every day of their lives may be actually more important than are beaches, if that should really have to be the choice. We also believe that the pollution which is produced in the air of New York City has by far a greater ultimate effect on the marine environment than do oil spills. This air pollution ultimately drifts seaward and deposits on water surfaces, affecting marine life on a continuous basis.

In short, the discovery of natural gas is far more important to the total environment than is any possible benefit from the prohibition of such drilling. Those truly interested in the ecology should be actively promoting the new discoveries of natural gas.

We can briefly summarize the position of Brooklyn Union by stating that we know there is a natural gas shortage. We are doing some unusual things to assure our present customers of adequate supply. Natural gas is essential for the environment of large metropolitan areas. We need to work with all those concerned with the ecology to obtain sufficient supply of this essential fuel. This is especially true in regard to a search for gas 30 to 300 miles off the east coast, where all the natural gas New York City would need for years to come may exist, within relatively easy reach.

## Introduction of Louis H. Roddis

*Clifton Daniel*

In introducing Mr. Griswold, I remarked that he has led a double life.

Louis H. Roddis, Jr., president of Consolidated Edison Company of New York, has led a double life in a different sense. Personally, I find it difficult enough to sustain one career. I am impressed and envious when I meet somebody who has been able to sustain two or three.

Mr. Roddis, born in Charleston, South Carolina, is an honor graduate of the United States Naval Academy, and holds a master's degree in naval architecture and marine engineering from M.I.T. While he was an officer in the Navy, Mr. Roddis helped develop a prototype ship propulsion reactor, and participated in the design of the first nuclear submarine, the Nautilus. He was chief assistant to Admiral Rickover, whom he calls a genius.

From the Navy he went to the Atomic Energy Commission, where he became deputy director of the Division of Reactor Development, and from there he went into a third career as president and later chairman of the board of Pennsylvania Electric Company. He joined Consolidated Edison, the nation's second largest privately owned utility, in 1969.

He is currently chairman of the executive committee of the New York Power Pool, and a consultant to several governmental agencies.

Mr. Roddis is also active in public service, charitable, industrial and trade organizations.

Louis H. Roddis, Jr.—for the electrical industry.

# ENERGIZING NEW YORK CITY

*Louis H. Roddis*

I am honored to speak at this important meeting. The job of energizing New York City is Con Edison's reason for existence. That in itself is a difficult job. To do it in environmentally acceptable ways makes the job more difficult. But it must be done.

Energizing New York City is a tough job because the City is so big. The equipment the job requires is awesome. Just one of our 40 distribution substations—the one which supplies electricity to customers in a portion of Queens—has a capacity exceeding the combined capacity of the four investor-owned utilities in the state of Rhode Island.

We are more than an electric company. In addition to the 32 billion kwhrs of electricity we sold in 1971, we sold 71 billion cubic feet of gas and 37 billion pounds of steam.

Today, under the streets of New York City and Westchester are enough gas mains and services to reach Tokyo, enough steam line to reach from here past Philadelphia, and enough of our electric cable to circle the earth three times.

The magnitude and complexity of the energy supply job, which these numbers illustrate, mean we must pay special attention to environmental protection.

The environmental problems which will be detailed at this afternoon's workshops demand protective measures. Con Edison has undertaken some important ones. Let me list some, and then let me turn to the role of research and development in meeting the challenge of protecting the environment. I would like also to take a realistic look at those methods of generating electricity which have been called the hopes of the future by some.

Switching to low-sulfur oil and stopping all coal burning are two ways to lessen air pollution and Con Edison has done both.

We now use nothing but low sulfur oil for electric production. We shut down our last coal burning plant on February 26, 1972.

There has been much recent talk of "total energy" systems to provide electricity which make use of the low temperature exhaust heat to provide space heating and air conditioning. Con Edison actually has the largest such system in the world. Much of the steam we distribute in lower Manhattan has already been used in so-called topping turbines to generate electricity before the exhaust steam is distributed. So through our steam system we get an "environmental dividend" from burning the same amount of fuel.

The energy supply for total energy needs is also important. Steam for heating and cooling is environmentally desirable. Since steam must be produced near its point of use, producing it with fuel burned at one point with a high stack is environmentally preferable to burning fuel in thousands of individual heating units.

But the most important environmental protection measure is one we can all adopt. We can all "Save-a-Watt." To the extent that people use less energy, we can do with fewer power plants. The frame of mind we all must accept is the one that says "Don't Waste." Use the energy you need for safe, healthful, productive living, but don't waste. Ask yourself if you *really* need more than one TV (see Figure 4). Ask yourself if you *really* need that air conditioner running all day when you're not home. Ask yourself: "Am I buying the most efficient appliance?" And when you answer that consider more than just the purchase price. Your electric bill will be lower if you buy an air conditioner that uses less energy.

You may say: "Okay, but what one person does won't really change things." Right. One person alone can't do much. But there are nearly ten million of us in this great city, and if only one among every ten of us "does something" it adds up to a lot. Each of us must do his part. Some 250,000 window type air conditioners are sold in Con Edison's service territory each year. Their electrical efficiency varies by a factor of more than two. If everyone bought the most efficient units instead of the least efficient cheap units, demand on the Con Edison system would be about 100,000 kw less this summer.

We have recently completed our 20-Year Advance Program —a look forward to 1990. This program is based on our projections and plans for the long-term future.

**Every "yes" answer helps to protect the environment, ease power emergencies, and reduce your bill.**

If you pass the save-a-watt test, you'll conserve electricity. You may also breathe a little easier in three important ways.

One way involves the quality of the air you breathe. Most electricity can't be produced now without some air pollution—although electric plants are by no means the worst offenders. To protect the environment, it's wise to conserve *all* forms of energy *all* year around.

There's another way you'll breathe easier if everybody saves a watt. There'll be less chance of power shortages this summer, when demand for electricity is highest. New plants have been delayed for reasons beyond our control. That means reserves are low, and we must rely more on older plants.

There's a third way, too. You'll save money on your electric bill.

The save-a-watt test in the next column points out 10 significant ways to avoid wasting electricity. Take it and pass it.

You'll help protect the environment.

You'll help reduce the risk of power shortages this summer.

And you'll save money on your electric bill.

Con Edison conserve energy

Figure 4. "Save-a-Watt" Commercial.
(Courtesy of Consolidated Edison Company)

An important assumption in the plan is that our peak load will grow by about 450,000 kw a year. This load growth appears to be a continuing one. It responds much more to the economy and growth of New York City than to our efforts to discourage unwise use of electricy.

Where will this energy go? Much of it will go for improvements in mass transit, sewage treatment, solid waste disposal, and many other environmental improvements. Much of it will be needed to sustain the huge construction projects needed to rebuild our cities. As the Reverend Leon Sullivan recently told a Federal Trade Commission meeting, rebuilding our inner cities is one of the greatest ways to halt urban decay, curb unemployment, and stimulate minority business.

One more example of the social benefits of energy use will suffice. The Waverly Center of the city's Department of Social Services is on 14th Street, about two blocks from my office. The lines start forming outside before 8:00 a.m. every morning. Those people are part of the more than 1¼ million of our fellow New Yorkers who need and receive public assistance. The number of persons *dependent* on that assistance is surely twice that. If we are ever to make any meaningful improvement in their quality of life, we will need more energy—more energy for jobs, for adequate housing, for a higher standard of living. As the celebrated blackout of 1965 so vividly showed us, whether our grandparents could live happily without electricty is not important. What is important is that we can't. But just as important is that we develop the environmental ethic that says "use what you really need (not what you think you need) and don't waste."

The time frame considered at this meeting goes to 1985. So let me look at what the future energy sources have in store for us. We must continue to energize New York City and present methods all make some adverse impact on the environment. So, we must avail ourselves of improved energy production techniques as they become available. More importantly, we need to hasten the day when new methods become realities. That takes R & D, and lots of it.

While there is no way today to generate electricity without some environmental impact, there are future energy sources, however, that may represent a substantial improvement in the way we produce and transmit electricity. Consequently, Con Edison has its own research and development program and supports industry-wide R & D efforts. Let me make three points relative to short and long term energy needs.

First, we have no choice but to meet today's demands for energy with today's technology. It takes so long to develop new technology—and then to build plants to use the new developments—that demands for energy even into the 1980's must be met chiefly with generating techniques available today. The new technologies cannot be developed to full application any sooner. Switching over night just is not possible. This is why the allocation of available fossil fuels is so critical. For example, natural gas is a diminishing, irreplaceable natural resource that many would preserve for what they perceive to be "higher end uses." But today's damaged environment does not always allow the long view. We can't hold our breath very long. So how we use the fuels we have is determined by what the facts of the particular case are, and the facts in the case of New York City are unique. We feel that the use of gas as a fuel for our steam system supplying lower Manhattan is a desirable use of gas just as it would be for any individual building heating use.

Second, future energy sources that have achieved scientific feasibility in a laboratory or pilot plant may still be a long way from commercial availability.

Third, even if commercial availability has been achieved, public acceptance of something new is far from automatic. This makes the direct public benefits of new technology even harder to achieve. One example will illustrate the time scale problems.

In 1942, scientists in Chicago demonstrated that controlled nuclear fission, the process driving nuclear power plants today, had achieved scientific feasibility. In 1953, the prototype of the Nautilus nuclear submarine, which I was privileged to help develop, demonstrated that the process had achieved engineering feasibility. Useful amounts of energy were produced. That was 19 years ago, and we certainly do not have universal public acceptance of nuclear power yet today, nor is any appreciable fraction of the country's energy today, thirty years after the invention, produced from nuclear sources.

With these admonitions before us, let me comment on some future energy sources.

Sources of energy can be divided into three categories: solar, earth, and nuclear.

## SOLAR—PRESENT USES

Solar sources include our well-known fossil fuels—coal, oil, and natural gas. All are derived from plants or animals that millions of years ago stored the sun's energy. Time and other processes changed them into the forms we know now. Con Edison no longer burns coal and is now largely dependent on oil.

Also included as a solar source is water power. The sun evaporates water and lets it be rained into higher elevations. From there it is available for harnessing by hydroelectric dams.

Combustion of the sun-produced fossil fuels could be improved environmentally in three ways. These are (1) a device called a fuel cell; (2) a process called magnetohydrodynamics (see Figure 5); and (3) a process to make natural gas, presently a desirable but very scarce fuel, from coal. All of these appear promising for development in the next 10 to 20 years. Con Edison supports research and development in all three areas.

An improved present-day use of the sun's heat would be to design buildings differently. The enormous glass houses of lower and midtown Manhattan, with their sealed windows, may be attractive but many are poorly designed to use or avoid the sun's heat with the seasons. More imaginative architecture could give us buildings that are easier to cool in summer and to heat in winter. The effect would be to reduce electric demand and thus conserve resources. This possibility is a here and now thing that building designers should use.

Coal, oil, natural gas, water power, and nuclear fission are the five common energy sources today. The drive to develop new sources comes from two things: the rapid dwindling of these existing sources, and the environmental disadvantages each source presents.

## SOLAR—FUTURE USES

The sun, representing as it does an almost infinite power source, certainly will be studied harder in the future. One day new ways to use its energy may come to fruition. So let us consider the future uses of solar power.

One solar source is wind power—air currents set in motion by the sun's energy. As recently as the late 1950's, about 300,000 windmills and 100,000 wind-driven electric plants were being used. They were only able to generate about 300,000,000 kwhrs of electricity, though. Last year Con Edison alone generated over 100 times that amount, the equivalent of 30,000,000 windmills and 10,000,000 wind-driven electric plants. This does not appear to be an energy source of any adequacy.

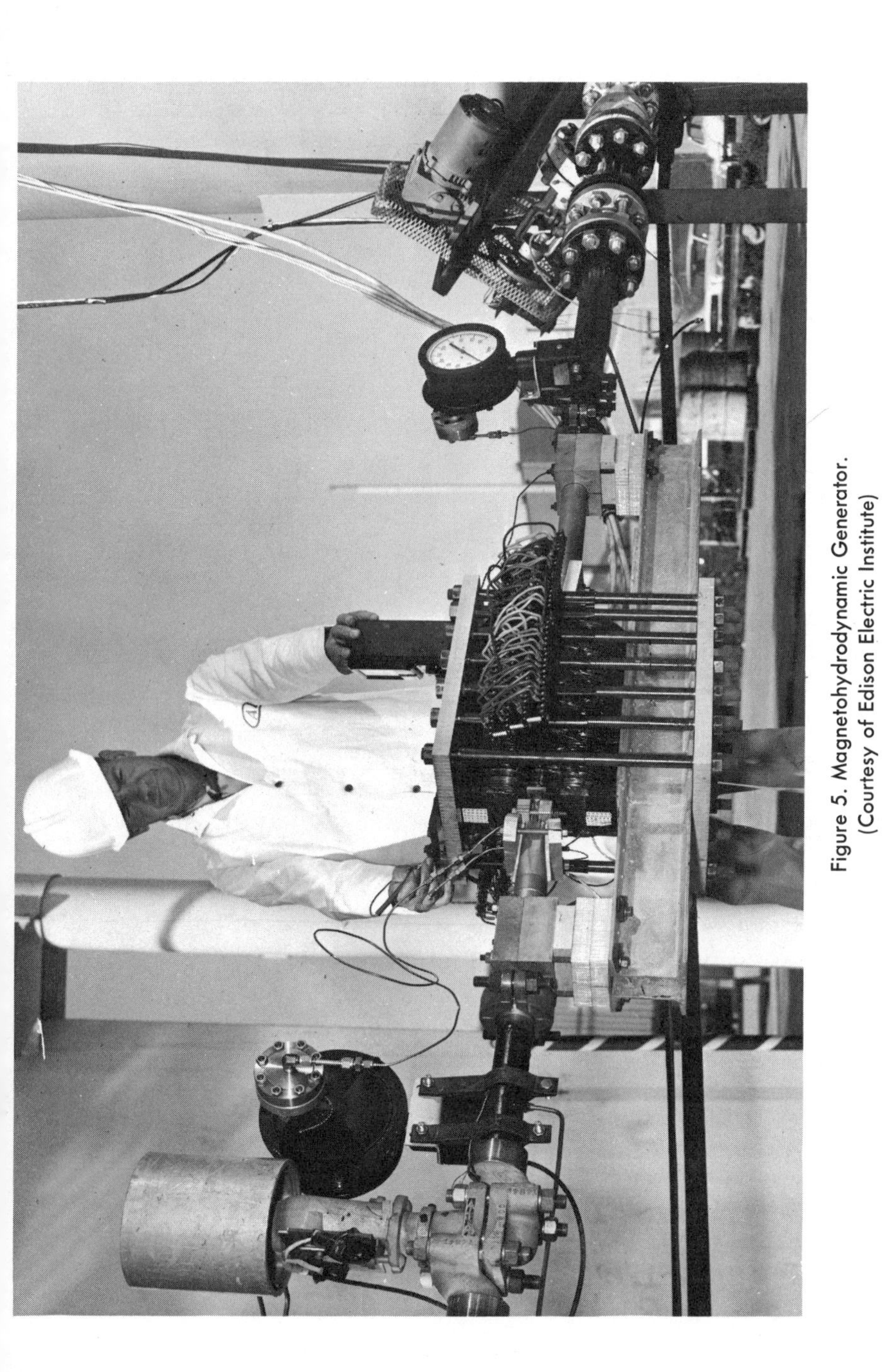

Figure 5. Magnetohydrodynamic Generator.
(Courtesy of Edison Electric Institute)

The sun's energy working through the process of photosynthesis can be used to produce green crops varying from algae to trees. These in turn can be refined into fuels such as methanol (wood alcohol). Such fuels now are expensive but some day we may see "Energy Farms" producing fuels from sunlight this way.

The temperature difference between cold and warm ocean currents, which is derived from the sun's energy, could be a source of useful energy. There are many problems to be solved, however. The process to use this temperature difference is extremely tentative, expensive, and by no means trouble free. This also does not appear to be a likely prospect in the near future.

The use of solar energy we hear about the most is the direct conversion process used in space satellites. A large square plate collects the sun's energy and converts it directly to electricity. The problem in using this approach to supply New York City is the size of the components and the effort and cost of building them. Technically, the process is certainly feasible on a small scale. But to supply New York City would require two collector plates in space. At the present level of technology, each plate would be five miles on a side and would be oriented in orbit 25,000 miles in space. These would convert solar energy to microwave electric energy which could penetrate clouds. The microwave beam receiver on land would be six miles on a side, or more than 1½ times the area of Manhattan! At present production costs, the two plates would cost about $800 billion. Assuming great improvements in cost and efficiency, we might lower the cost to a few billion. This excludes the cost of developing and building the necessary space shuttles to orbit the collector plate, which is also in the billions. There are also the problems of keeping people away from the microwave beam that transmits power from space, and of keeping the beam from wandering off target. This does not appear to be too promising a large scale approach in the near term future.

We may also consider using the desert areas in parts of the world as solar energy receivers. The large land areas required are available and the sun shines a great portion of the time. These are two advantages that clearly do not exist in New York, with our weather and congestion. The problems to be overcome in desert application are transmission of the electricity to the centers, where the people are, and of the conversion techniques to make electric energy continously. Using the sun's energy to heat some fluid could allow a sort of "energy storage." This would help overcome the obvious problem of having no sunlight—and thus no electricity—at night. These approaches are now receiving attention in parts of the world where they are appropriate.

## EARTH

Moving now to earth sources, we consider the two potential sources of energy derived from the earth's composition and motion. These are geothermal power and tidal power.

Geothermal power relies on steam produced naturally beneath the earth's surfice. Known sources of this steam have been used where they exist, but they are very scarce and not very large. Recent interest has indicated that sources of geothermal steam may be more widespread than previously suspected. They may be less accessible, and could require extensive drilling to be tapped. We and others are beginning to take a look at this possible energy source, though it too has environmental problems.

Tidal power suffers the same drawbacks. It can only be tapped where tides rise and fall great amounts. No such source exists convenient to New York City. The Passamaquoddy-Bay of Fundy region near Nova Scotia has been studied many times as a possible tidal power scheme but the economics simply do not work out favorably. The most recent study by the Canadian Government confirms this.

## NUCLEAR

The third source of energy is nuclear. Nuclear fission, the splitting of atoms, is what drives today's nuclear power plants. Today's plants are inefficient users of uranium but the fission process they use is the only-one we have developed to commercial availability (see Figure 6).

We are continuing to explore new concepts in nuclear plant siting, such as plants located on barges anchored from three to ten miles out to sea. This idea is not universally accepted, however. One house of the New Jersey legislature voted 65-0 in opposition to specific barge-mounted nuclear siting plans of Public Service Electric and Gas Company. In addition, the legal aspects of going beyond the three mile limit are not clear.

A major improvement in the fission process is to use nuclear power plants called breeder reactors. A breeder reactor produces more fuel than it consumes. This would obviously be an efficient way to conserve resources. Con Edison actively supports the national program of research and development in breeder reactors.

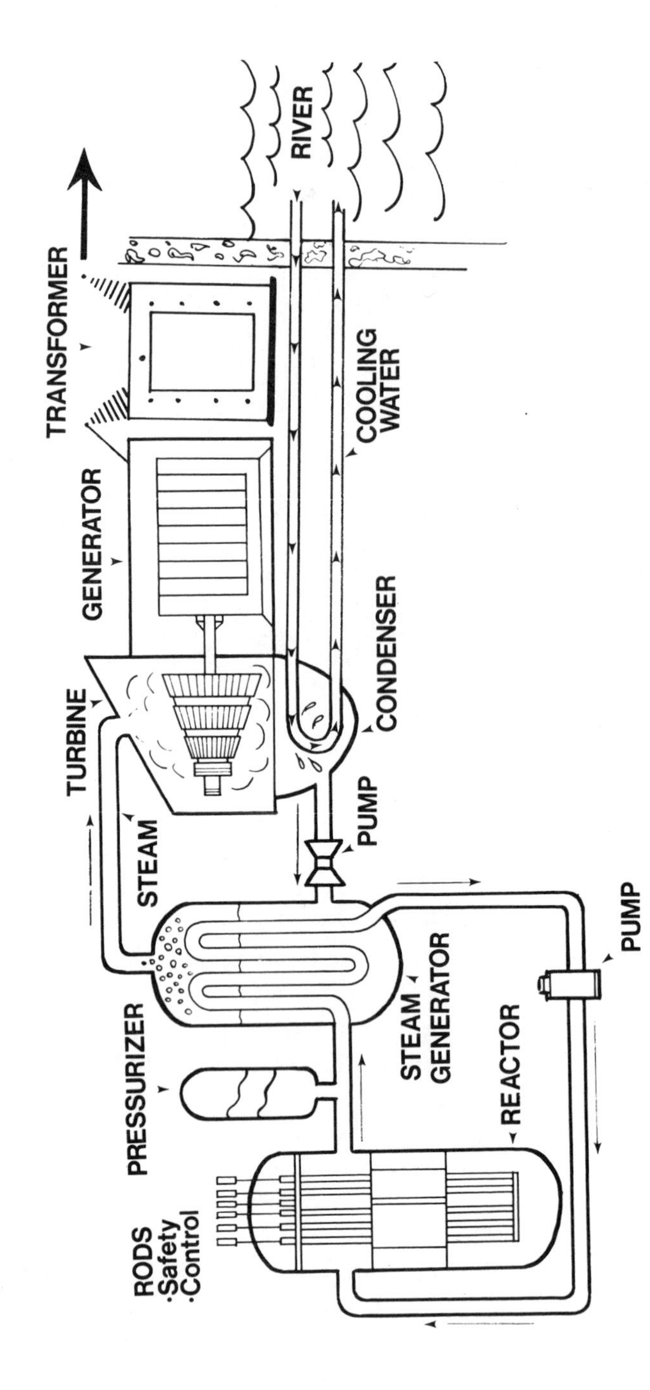


Figure 6. Pressurized Water Reactor.

We have already expended over $1,250,000 on breeder reactor research and are contributing nearly $8,000,000 in the next ten years towards this program, which will produce a national demonstration plant early in the 1980's.

The ultimate nuclear source seems to be fusion. This is the energy source of the stars, including our own sun. Fusion is the joining of atoms, in contrast to fission, which is the splitting of atoms. Today controlled fusion is still many, many years away. The biggest problem in controlling fusion is the high temperature required: about 100,000,000 degrees. Finding a way to achieve that temperature is no easy task, but there has been much work done on this over the last twenty years. Con Edison and other utilities are contributing to increasing research in fusion, because it is our best hope for a relatively inexhaustible, relatively pollution-free form of energy. In the ultimate development of fusion, the fuel, readily available from water, could last us for centuries, but it will be several decades or more before it is available for commercial use.

Let me conclude with this. The job in the years ahead must be to supply the energy the city needs to grow, to remain healthy, and to improve the quality of life for all New Yorkers. This is especially true for those people standing outside the Waverly Center and the others like them. The direction we take, though, in doing this job, must be along the road of reason and restraint. Unlimited growth is just as unacceptable as zero growth. The environmental ethic that leads all of us to use all forms of energy wisely and not wastefully must become second nature to us all.

## Introduction of Robert N. Rickles

*Clifton Daniel*

Our next speaker, Robert N. Rickles, "is a nut and a damn fool." Those, I hasten to say, are not my words, but the words of Mayor Lindsay of New York. Not many weeks after they were uttered, Mr. Rickles, not surprisingly, resigned from Mayor Lindsay's administration. He was Commissioner of Air Resources.

He had aroused the Mayor's anger by opposing the New York State Transportation Bond Issue on the November election last year, because he thought the bond issue was oriented too much toward highway construction. He was the only major city official to oppose the Mayor on that issue.

Since resigning from the city government, Dr. Rickles has organized and is now director of the Institute for Public Transportation, a non-profit group devoted to the conviction that transportation is the determining environmental issue.

Dr. Rickles—his doctorate is in chemical engineering—has written seven books and numerous articles. Before joining the city administration he had his own environmental consulting firm, and has been engaged in almost every environmental battle in and around New York in recent years. He works in New York but in the evenings and on weekends he breaths the cleaner air of Stamford, Connecticut.

Dr. Rickles—speaking on behalf of environmental protection.

# CORPORATIONS IN AN ENVIRONMENTAL AGE

*Robert N. Rickles*

The world in which we live is not shaped as much by philosophy as by the pragmatic actions of people aiming to accomplish certain objectives. But the musing of men helps establish the consensus for those objectives which guide the world.

After a decade of opposing certain specific manifestations of broad economic policies, environmentalists are now proposing new objectives to guide us, for we have reached the age where the consequences of the technological changes of the recent past threaten not merely our comforts, but our very existence. Living in an age of unparalled consumption, unlimited growth, and continuing technological inventiveness, it is difficult to visualize a day just over the horizon when man will face not simply intolerable pollution but also a shortage of the raw materials upon which our technological society depends.

Such are the prophecies of two studies, which are made even more frightening by their own dependence upon modern technological forecasting techniques. The studies, by Forrester at MIT sponsored by Club De Roma, and the joint statement of England's most respected scientists, indicate that unless growth is halted within the next two decades, the world faces an intolerable crunch between a steadily deteriorating environment and a permanent shortage of raw materials resulting finally, if not in the elimination of man, at least in the end of society as we know it. The two reports forecast that these events will take place within the next hundred years, within the lifetime of my own grandchildren! While the exact date is certainly in doubt, the continued use and dilution of mineral and energy sources certainly means the uncontestable result as forecast by Forrester at a date far too early to provide comfort for any of us. It is not simply the growth of the developed countries which puts a squeeze on these resources, it is the desire of the undeveloped three-quarters of the world's people who will, in the remainder of this century, demand to live as we do now, which will put the real strain upon these limited supplies.

But growth, Treasury Secretary Connally tells us, is necessary to end poverty for some 40,000,000 Americans. Growth is ingrained in the American character. It is the rare institution, be it corporation, union, or government agency, that is not planning for growth, proclaiming proudly its growth or bemoaning its lack thereof. American leaders have said so often, "to live is to grow, to cease growing is to die." Yet growth has not produced a reduction in the number of poor in this country. In the face of the greatest growth surge of all time, the number of people on welfare has increased to a record number. With all of this growth, the amount of pollution in many areas of the world has reached almost intolerable levels, in some cases shortening the lives of ten of thousands. The growth era in America has seen the decay and wasting of our cities, the asphalting over of our countryside, and the destruction of the peace and beauty that was once available to even the most humble American. Work, which once was a source of pride and envy, reflected in both the product and the workman, has in the name of productivity become a meaningless and non-fulfilling time consumer for the average worker.

It almost seems as though growth causes, as side effects, enough social problems to absorb the extra wealth it produces, and produces few, if any, real benefits to society.

Corporations are, because of the nature of their structure, their internally channelled capital and the human competitive spirit, the prime generators of growth, and especially of the mindless growth which provides goods of dubious need for those who already have too much and services which are as often wasteful as beneficial.

In this manner the corporation finds itself in direct confrontation with the environmental age and environmentalists. The need to grow, to generate profits, and the seeming inability to change direction is often cited by environmentalists demanding the end of the profit motive and corporations. It should be quickly added that these environmentalists also note that communism is nothing more than state capitalism with the same built-in demands for growth.

Can it then be said that the private corporation and the environmentalist movement are on a collision course? I believe so, but I am not at all certain that this is necessary. I believe that we can make changes both in the objectives of our society and the way in which we achieve these goals over a time frame which will permit a "rolling adjustment" for both corporations and individuals. Again, I want to repeat that the need for changes in our social objectives and pragmatic efforts are not to satisfy environmentalists, but to prevent the all too certain occurrence of intolerable pollution and resource exhaustion if current practices continue in the future.

What then are those steps that are necessary to move from today's world to the steady state society visualized by John Stuart Mills a hundred years ago?

They are:

### 1. THE END TO GROWTH OF PRODUCTIVITY

Increasing productivity has been the hallmark of our technological age. It is the essence of the appeal of the computer and automation in general. Now we have to recognize that it is this very rate of productivity which is choking us. Increasing productivity automatically produces an incrementally greater increase in consumption. But most important, it leads to underemployment, waste, and a diminution in human spirit. We must begin to move again towards more meaningful, dignifying labor. After all, work is only valuable if it produces pride, dignity, and happiness in the worker as well as products and services. If it does not produce this, then another human will die never having been part of the society. We must restore pride in working and in workmanship. The terrible example of the GM Vega Plant at Lordstown, Ohio, shows that the move towards increased productivity will produce a destruction of the human spirit. If we believe that most people need to work, then we must avoid productivity increases, because in a world of decreasing raw materials, increasing productivity can only mean more machines, massive underemployment, and meaningless jobs for most.

### 2. ACHIEVEMENT OF ZERO POPULATION GROWTH

Nothing is clearer than the fact that increases in population, especially in undeveloped countries, bear the seeds of future

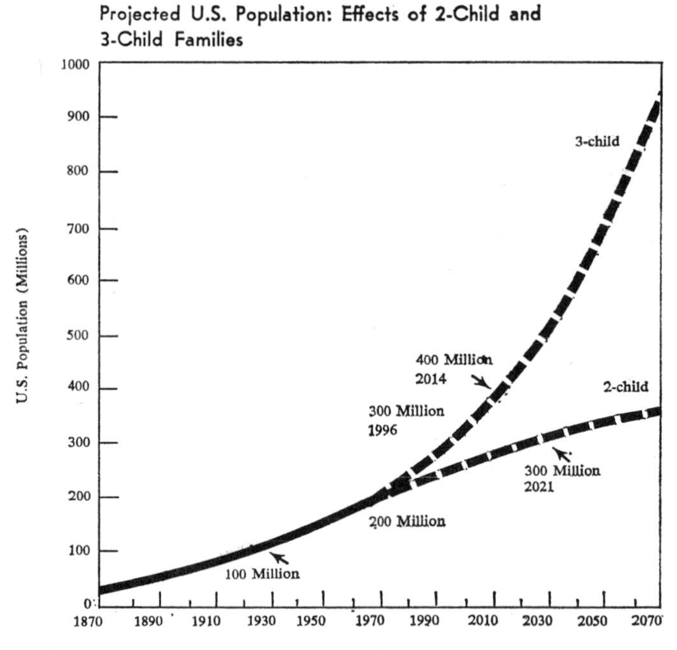


SOURCE: Commissiön on Population Growth and the American Future.

Figure 7.

doom. If we cannot achieve a zero population growth in the world, not merely in this country (see Figure 7), then any hope for achieving a steady state society is nonexistent. We may be already perilously close to a reasonable maximum population. The English study puts the number based on per capita calorie intake at 3.5 billion. Today, as no time in the past, the goal of zero growth is in sight. In almost every country on earth, birth rates are declining. In the U.S., we are hovering just above the zero growth level. Further education as well as extension of the right of abortion and the development of better birth control methods for both males and females are urgently needed.

### 3. TECHNOLOGICAL INNOVATION

We have relied on the technologists to produce the current age. No one today can remember a time when science and technology could not respond to the needs of society for improvement. Yet to some degree science has failed us because we have asked science to solve false problems, and ignored the real ones that infest us. We should also be aware that the growth of knowledge is not a steady eternal phenomenon. It may be that we will enter an age of relative technological sterility. For example, the number of new concepts arising in industrial laboratories has been steadily dropping over the last decade despite an increase in R & D expenditures. However, we should now focus on our real problems, and see if technology can solve them under special programs. If I had to choose a single technological problem, it would be that of energy. I would recommend that this country undertake, on NASA basis, a $10 billion/year program over the next two decades to develop fusion power. I do not believe that there exists any other alternative to the avoidance of an energy squeeze with the exhaustion of both fuels and the environment due to the forecasted expanded production of energy. The President should ask for and the Congress should fund such a program at once.

## 4. GOVERNMENT INTERVENTION

The profit motive need not be eliminated. It can operate just as efficiently for socially useful purposes, like mass transit, better housing, and better health services, as it now does to build automobiles, commercial real estate and ABM's. To suggest that this means improper government intervention is to ignore the reality of present government intervention which rewards highway builders and defense contractors. The government need only change its priorities to make the profit motive work for more socially useful purposes.

Government must spend more thought in developing policies consistent with the environmental age. For example, the recent decision of the President to drop the 7% excise tax on autos to stimulate the economy was a monumental mistake because it produced growth in an industry that is specially destructive of the environment and wasteful of resources. To stimulate the economy, the government should have spent the $2 billion on Public Transportation equipment.

If the government wishes to reduce the waste of raw materials, it must use its regulatory and taxing power to assist in recycling. Depletion allowances encourage the use of newly mined resources rather than recycled ones. Transportation rates make it cheaper to ship ironore to Cleveland from Duluth than to ship scrap steel from Toledo.

## 5. THE END TO GOVERNMENT GROWTH

Government today lives largely for its own growth, thereby

necessitating the growth of the overall economy which feeds it. So long as government must live on the taxes from the incomes of individuals and corporations, it acts on its own internal growth interests to continually increase those incomes. Less government, rather than more government must accompany reductions in private productivity. Exercising of governmental functions on the lowest scale possible should be an objective.

Perhaps it will be possible to create a society in which, as Pericles said, "We are prevented from doing wrong by respect for authority and for the laws, having an especial regard for those unwritten laws which bring upon the transgressor of them the reprobation of the general sentiment"; or, as Solon said when asked the name of the best policed city, "the city where all the citizens, whether they have suffered injury or not, equally pursue and punish injustice."

## 6. REDISTRIBUTION OF INCOME

Obviously we must address ourselves to the presence of tens of millions of poor in this country, and billions of poor throughout the world, and redress their grievances before any total halt to growth may occur. This will require real and not cosmetic changes in our society, including the adoption of a substantial system of income redistribution in this country.

## 7. CHANGES IN TECHNIQUES

We have in this country adopted techniques that are extraordinarily out of phase with the conservation of resources and the environment, as well as wasteful of money. For example, we have adopted a transportation system which centers upon the motor vehicle. This system, besides its well-known adverse impact upon the human hearing and respiratory systems, also kills 50,000 Americans each year, and maims two million more. But equally important, it consumes three hundred billion dollars, 30% of our gross national product, and substantial amounts of our rubber, energy and metal supplies. It is six times as wasteful of energy as alternative public transit. It is imperative that we move toward public transit and away from the automobile. It is imperative that we examine each segment of our national life and move away from resource wasting techniques towards resource conserving ones.

## 8. WORLD WIDE DIRECTION

Finally, we must, first by example, teach the underdeveloped world to avoid the tragic examples of our experience. We must not export consumerism and materialism as Henry Ford II is now attempting to do in building Ford plants in the Near East. The fate of the corporation in the environmental state rests with its self-proclaimed ability to adapt to new environments. If it can do so, if it can adapt to a steady state society, then it will survive; but if it cannot, then the fate of the dinosaur awaits it.

As for the individual, the 19th Century philosopher who was the most articulate defender of individual liberty, John Stuart Mills, also advocated an economically stable society. We have chosen to build one part of his writings into every school child's experience, but denied the truth of the stationary state. His observations are as valid as they were a hundred years ago. I leave them with you:

"I cannot . . . regard the stationary state of capital and wealth with the unaffected aversion so generally manifested towards it by political economists of the old school. I am inclined to believe that it would be, on the whole, a very considerable improvement on our present condition. I confess I am not charmed with the ideal of life held out by those who think that the normal state of human beings is that of struggling to get on; that the trampling, crushing, elbowing, and treading on each other's heels which forms the existing type of social life, are the most desirable lot of human kind . . . the northern and middle states of America are a specimen of this stage of civilisation in very favourable circumstances: and all that these advantages seem to have yet done for them . . . is that the life of the whole of one sex is devoted to dollar hunting, and of the other to breeding dollar-hunters.

"I know not why it should be a matter of congratulation that persons who are already richer than anyone needs to be, should have doubled their means of consuming things which give little or no pleasure except as representative of wealth . . . it is only in the backward countries of the world that increased production is still an important object. In those most advanced, what is economically needed is a better distribution, of which one indispensable means is a stricter restraint on population . . . the density of population necessary to enable mankind to obtain, in the greatest degree, all the advantages both of cooperation and of social intercourse, has, in all the most populous countries, been attained . . . it is not good for a man to be kept perforce at all times in the presence of his species . . . nor is there much satisfaction in contemplating a world with nothing left to the spontaneous activity of nature . . . if the earth must lose that great portion of its pleasantness which it owes to things that the unlimited increase of wealth and population would extirpate from it, for the mere purpose of enabling it to support a larger population, I sincerely hope, for the sake of posterity, that they will be content to be stationary, long before necessity compels them to it.

"It is scarcely necessary to remark that a stationary condition of capital and population implies no stationary state of human improvement. There would be as much scope as ever for all kinds of mental culture, and moral and social progress: as much room for improving the art of living and much more likelihood of it being improved, when minds cease to be engrossed by the art of getting on."*

*John Stuart Mill, *Principles of Political Economy,* Vol. II. London, John W. Parker, 1857. (In "A Blueprint for Survival—The Goal", The Ecologist, Vol. 2, No. 1, January 1972).

## Introduction of David Sive

*Clifton Daniel*

David Sive is listed in *Who's Who* as a lawyer. You might as well identify Richard M. Nixon as a lawyer. That doesn't being to tell the story.

Mr. Sive is indeed a lawyer and a very active and prominent one. He has also been active and prominent in Democratic politics. But he could more fully be described as an outstanding environmentalist and, that, of course, is the capacity in which he appears here today.

Listen to his credentials: Executive Director of the New York Constitutional Convention Committee on Natural Resources and Agriculture; member of the Board of Directors of the Scenic Hudson Preservation Conference; National Audubon Society; Wilderness Society; Rockland County and Delaware County conservation associations; trustee of the Natural Resources Defense Council; member of the Legal Advisory Committee to the Council on Environmental Quality; member of the New York City Mayor's Committee on the Environment; a director and member of the Executive Committee of Friends of the Earth; and legal advisor of the Sierra Club.

David Sive—representing the environmentalists.

# SYNOPSIS OF REMARKS

*David Sive*

Addressing myself to the place of lawyers in the environmental protection-energy production problem, I find cause to dwell on a perspective first voiced from the bench of a federal court a few years ago. A somewhat harried plaintiff was questioned concerning the identity of his counsel, to which matter of representation he replied, "I stand here in the face of all mankind, in righteousness and truth; with God as my witness, my advocate and my judge." "Yes, and that's all very well, I'm sure," replied the judge, "But, who represents you *locally?*"

And that is perhaps a point, that we lawyers are, in certain respects, at least for environmental protection matters, cast in the roles of "local representation" for what are really somewhat loftier matters than just a single power plant siting controversy or just one rate case.

This conference involves the problems of integrating demands for energy and protection of environmental resources; in the whole recent history of the environmental movement that has been a problem at the center of many major controversies.

Beginning with the classic and still ongoing dispute about the pumped storage power plant at Storm King Mountain, the central theme for environmentalists, for whom I am privileged to speak this morning, stated simply is this: "environmental considerations and considerations for energy supply rank equally in importance."

In the working out of this balance, advocates for both sides must retain a clear sense of their role, and argue strongly for it. The older model of "wise men making decisions alone" is inadequate for full development of the case in matters of modern technology and complex interrelations with the economy, public safety and security. The fostering of a dialog, an active exchange of conflicting ideas, is essential to retaining and to clarifying the interests of both parties in a public record. But conflict in an open, advocate forum need not imply ill-will or intolerence on either side; let me re-emphasize the value of "civility," as it has been called, as the preservative of constructive disagreement. Embodied in the value of open discussion is the public's right to discovery of essential information in a timely manner, to cross-examine and check expertise, to balance the evidence against the criteria of judgement.

The New York State power plant siting legislation recently enacted by the legislature and shortly to be signed by the Governor embodies the balancing principle, though its significance might not be readily apparent to environmentalists. The point is that just as on some occasions environmental needs must be subordinated to needs for power, on other occasions alleged needs for power must be subordinated to environmental protection.

The whole subject of electricity and the environment—most immediate of the New York City energy problems—is under intensive study by a number of agencies, one of which is the Special Committee on Electricity and the Environment of the Association of the Bar of the City of New York. I am a member of that committee, and by coincidence its preliminary draft report has been made public this morning.

One of our fundamental recommendations, consistent with the position of most environmentalists, is that the right to contest need for and siting of a powerplant in an adversary proceeding should be preserved. The Committee report, furthermore, attempts to divide controversies involved in such proceedings into generic and particular-plant issues. This we hope will preserve the adversary process to litigate the alleged need for power, for example, but not require that it be relitigated in each siting proceeding.

Another recommendation is that a Federal Agency allocates to several regions of the country the duty to produce the amounts of power determined from time to time by the Congress. This would mean that just as a locality within a state may not rule plant facilities out of its area so a state may not rule itself out of any obligation to use any of its area to supply the power needs of a large metropolitan area.

Other recommendations deal with the problem, or alleged problem, of delay in siting proceedings arising out of environmentalists and other citizens groups' intervention. The essential procedural recommendation is that the public participation start much earlier than it has, and that intervenors be granted full rights of discovery just as the ordinary litigant in a civil court action. The idea is to hasten the process overall but not by squeezing the hearing itself. If the public part could begin during the preparation rather than being pressed into the weeks following the license hearing notice, the time could be used to narrow and compromise the issues, and bring to the hearing a tighter agenda without sacrifice of essential questions in the public interest.

In fact, it has been noted by utility and federal agency alike that delay in power plant siting caused by environmentalists is not significant when measured against other causes like strikes, late delivery of major components, and technological refits.

One further point: environmentalists see power demands as integral to the whole problem of growth. It is no longer unpatriotic or neurotic to question the religion of growth. We have asserted, in both political and judicial contexts, and will continue to assert, our beliefs that government should use their powers to aid their citizens in securing what truly contributes to what Thoreau called the "elevation of mankind." That elevation may be secured as much by turning off some projects as by multiplying them. We are not against growth per se; we are against the blind pursuit of growth.

## Introduction of Kerryn King

*Clifton Daniel*

Lawyers, chemical engineers, naval architects and environmentalists are fine, but I have to confess a closer personal affinity for our next speaker—Kerryn King, senior vice president in charge of Texaco's Worldwide Sales, Public Relations and Personnel.

Born in Dallas, he started out with a degree in journalism from Southern Methodist University in 1939—he got an honorary doctorate in 1966—but like so many journalism graduates, he was diverted from journalism into public relations, and I would guess he is none the poorer for it.

After working in the electric utility and aircraft industries, he joined the international public relations firm of Hill & Knowlton in New York, and became a senior vice president there in 1952. He went to Texaco as director of public relations in 1953, and rapidly moved up the ladder until he was elected to his present position last year.

Along the way he worked not only in public relations but, what is surprising for a P.R. man, he also held several operating jobs at Texaco. From 1963 to 1965, for example, he was in charge of the company's Latin American operations.

Kerryn King—for the oil industry.

# ADDRESS

*Kerryn King, Sr.*

One of the great difficulties encountered in any forum assembled to consider the energy question is the multiplicity of viewpoints. Most of you, I am sure, have attended many sessions on this subject in the past. And I am equally sure that the net results of your deliberations were far from satisfactory.

I don't mean to minimize the need for information. But our purpose here is to determine courses of action. On the one hand, we need to avoid the kind of panic-inspired regulation and legislation that has been repeatedly proposed and sometimes adopted in recent years. On the other hand, we need to forgo the fruitless rhetoric that has engulfed the energy-environment issue and get on with the business of programming and planning our society's future progress on a sound, correct, and rational basis.

At this point, there seems to be agreement among the various involved parties only at the loftiest levels. We all want what is best for this nation, this state, and this city—but we do not yet agree on what "best" is.

My purpose in being here is to participate in what I hope will be a dialogue, not a debate. I offer you the facts, figures, and the position of the petroleum industry, not to rebut the views of others, but rather because this information, so vital to the issue, is all too often ignored.

Others are asking, with great justification: What would life be like without clean air and clean water? Without open, unspoiled areas? Without wildlife and wilderness? What would life be like with populations in the multi-billions? And with industrial activity pushed to the ultimate extremes necessary to support runaway growth?

I ask you to consider also, what the world would be like without adequate supplies of energy.

First, we must recognize that this is a very real possibility.

In 1970, oil and gas represented 77% of all the energy consumed in the U.S. By 1985, the requirement for these fuels probably will increase by 48%. Cumulative requirements for the 15-year period are expected to exceed 100 billion barrels of oil and more than 300 trillion cubic feet of gas. However, combined production of domestic oil and gas is now very close to maximum capacity, except for the giant Prudhoe Bay field on the North Slope of Alaska.

The reason why U.S. demand is expected to increase so much by 1985 is obvious: increased population and improved standards of living. Not so obvious, however, is a workable means of coping with the enormous energy demand problems that will be created by these two factors in the future.

Now, theoretically, there are negative solutions to the problem of future energy demand. We could opt for zero-population-growth and zero-life-improvement. Zero-population-growth is achieved when births do not exceed deaths. But quite apart from the moral and ethical problems involved, the zero-population approach is not a practical solution to the current problem. No population control program initiated today or tomorrow could have any significant effect on energy requirements now projected for 1985. The population age group 0-13 years is not, after all, a large consumer of energy. Our current projections are based primarily on the future neds of people who are now living.

Zero-life-improvement is a fascinating concept—but I wouldn't relish the prospect of running for political office with a plank like that in my platform. I'll wager that there aren't many people in this room who would happily accept the idea that their standards of living have risen as high as they should go. And this is a relatively affluent group. There are more than 100 million Americans out there whose living standards are below ours.

And what about the dreams and aspirations of those still living on the borderline of poverty and despair? It's not conceivable to me that this nation could advocate policies that would, in effect, say to its citizens, "You have gone as far as you can go."

I submit then, that we have no option. At least into the foreseeable future we must find ways to supply, not reduce, the projected energy demand. The question is, how to do it with a minimum of damage to our social, economic, and political values—including environmental values.

Here the cost-benefit factor must be applied to any possible future course of action.

In the not-too-distant past, the United States was able to produce all the oil and gas it needed from domestic sources. Last year this country was dependent on foreign sources for about 26% of our oil supplies (see Figure 8). Even assuming that Alaska's reserves are brought into production by 1976—but also assuming that U.S. production elsewhere continues to decline—this country could be 50% dependent on foreign oil by 1980 and close to 60% dependent by 1985.

Now the benefit of foreign oil imports is obvious: we need energy. But the cost factor is not closely enough examined by those who argue for removal of all controls. The price of foreign oil is admittedly lower than the price of domestic oil, but it is on the rise. Because domestic oil cannot compete against foreign oil in an open market, an unrestricted flow of foreign oil into the United States would destroy the domestic producing industry. And I said "destroy," not "shut down." Moreover, it would create a seller's market that would drive the price of foreign oil up and quickly cancel out any price advantage that might have temporarily existed.

And then there are the hidden costs:

The effect of uncontrolled imports on the U.S. balance-of-payments would be devastating. There would be massive unemployment created by the loss of jobs in the oil and related industries. Millions of U.S. investors would suffer great losses, and thousands of small businesses would go under. And last, but not least, this country would find itself dependent for its primary energy supplies on foreign nations that are not noted for their political stability and reliability.

Clearly, when we look at foreign oil the cost-benefit ratio is out of balance. What then are the alternatives?

Nuclear power would certainly help to relieve the pressure. And, believe me, the petroleum industry would welcome such a development. But I don't think I need remind you of the difficulties that the nuclear energy industry has encountered since its inception. Rising costs, technical problems, and fierce resistance to new nuclear power plants have prevented this industry from coming even close to its original goals. And even if nuclear

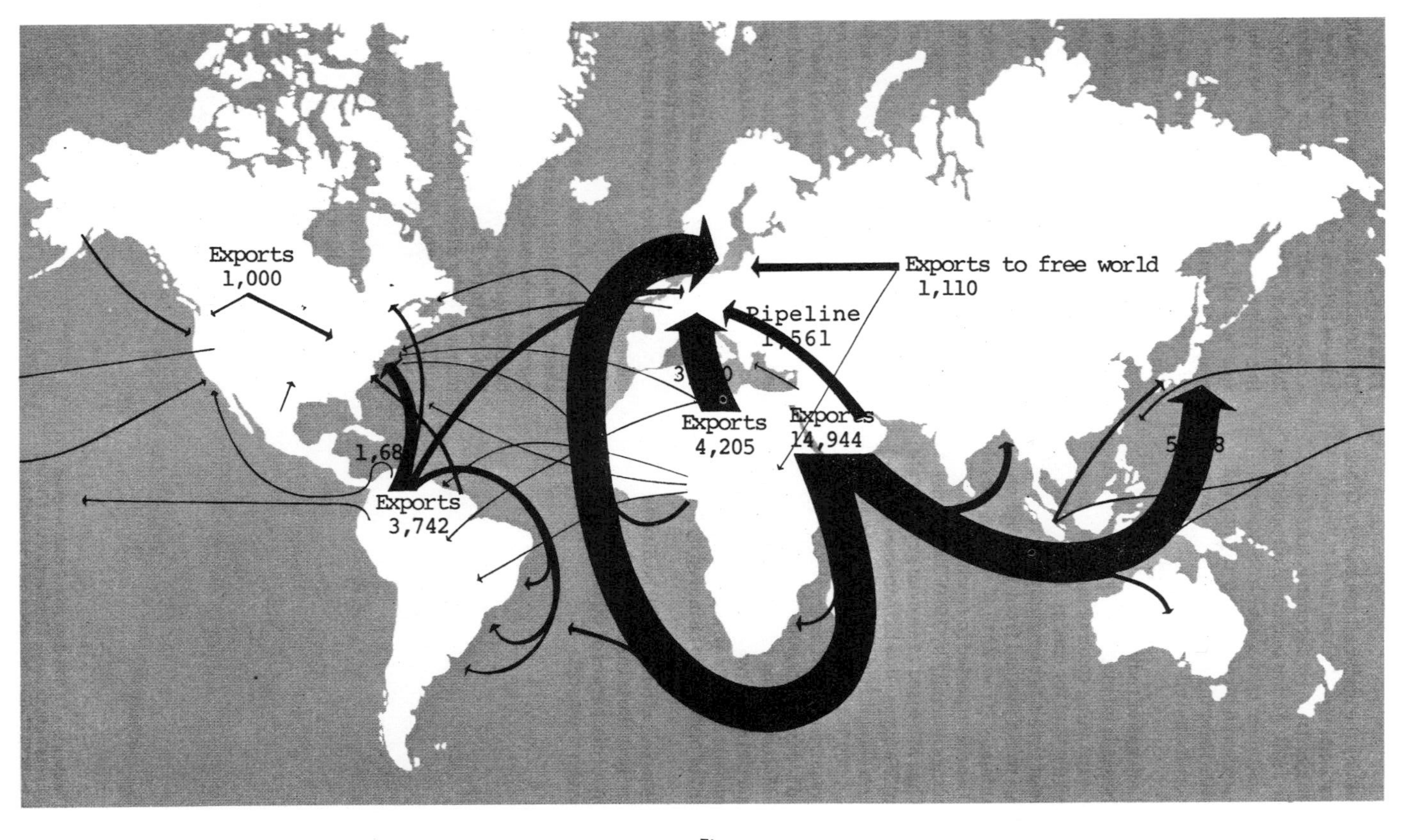
Exports
1,000
Exports to free world
1,110
Exports
4,205
Exports
14,944
Exports
3,742

Figure 8.

energy meets its present 1985 goal, this industry would be able to supply only 17% of the nation's total energy needs at that time.

There is some prospect of recovering a type of crude oil from enormous tar-sand deposits and of developing synthetic oil and gas from the huge coal and oil-shale deposits in this country. But there are problems here also. The bulk of U.S. tar sands are too deep to be mineable, and a practical method for recovering this crude without mining is not now available. Also, present technology does not permit us to make synthetic fuels at costs even approaching the costs of conventional fuels. Even if the technology is developed it will be many years before significant amounts of fuel are available from these sources. And there is the possibility that environmental restrictions will also block progress in this area.

What this all adds up to is a clear indication that the United States must do everything possible to develop additional domestic reserves of oil and gas in the years ahead. We simply must find enough supplies to keep our foreign dependency at a minimum until alternate energy sources become available in amounts sufficient to supply a significant portion of demand.

This means:

First, the petroleum industry must be given the freedom to search for additional domestic reserves of oil and gas wherever they are most likely to be found. Since the most promising onshore areas (see Figure 9) in the lower 48 states have already been thoroughly explored, we need to intensify the search in offshore areas and in Alaska. Environmental restrictions against drilling in such areas as offshore California and offshore Louisiana must be lifted, and we must accept the need of exploring promising areas on the East Coast Continental Shelf, as well. The cities and towns of the Northeast are among the nation's principal consumers of oil products, and they have the greatest need for increased supplies of clean-burning natural gas. With so much at stake, it would be highly irresponsible if the people of the Northeast were to refuse to do their part in helping to solve a serious national problem in which their regional interest is greater than any other section of the country.

Last week, Administration approval was given for construction of the Trans-Alaska Pipe Line. This is good news. But it is far from being a solution to all our energy problems. The field discovered at Prudhoe Bay is the largest ever found in the

Figure 9. Drilling Rig.
(Courtesy of Brooklyn Union Gas Company)

United States. But the estimated crude reserves in this field represent only about two years of U.S. oil consumption at current annual rates, and, therefore, the Prudhoe Bay Field will not free us from dangerous dependency on foreign oil. And it should be stressed that even if construction begins this year, Prudhoe Bay reserves will not be a significant source of supply before 1975 or 1976.

Nor do we have any assurance that Alaska contains additional undiscovered fields of comparable size. Whether Prudhoe Bay is a unique giant in an area with much smaller fields, or is only one of several giants, is a question that can be answered only by further drilling and development. Results in Canada's Mackensie Delta are encouraging. But essentially none of the required exploratory and development drilling can proceed until an answer is obtained to the question of how the crude can be transported to the "lower 48."

As equally urgent as the need to increase reserves is the need for additional transportation, storage, refining, and marketing facilities. There is not a single deepwater port in the United States that is now able to handle deliveries of imported crude oil by mammoth tankers. Without such ports, we are forced either to ship long distances in small tankers or to transfer cargo from mammoths to smaller vessels to supply our refineries. This not only increases costs, but it also increases tanker traffic.

We are all familiar with the complaints of the New England states about high prices of petroleum products in their marketing areas. But not everyone realizes that these same states have consistently blocked the construction of needed facilities within their borders. If various regions, such as New England, continue to reject oil exploration, port development, and refinery and nuclear power plant construction, and simultaneously enact legislation and regulations that unnecessarily restrict the use of available fuels, they must be prepared to accept the consequences—recurrent and increasing energy shortages.

This holds equally true for New York. Restrict the flow of energy into New York, and you make it colder in winter, hotter in summer, darker at night, and far less industrious.

The costs of unrealistic attitudes are measurable in dollars, as well. Even in the best of conditions, energy costs are going to go up. There are fewer sources available, and it is becoming increasingly more costly to supply. It is estimated that the petroleum industry will have to invest about one trillion dollars in additional facilities between now and 1990 to meet rising demand in the Free World. A good part of that sum will have to come from increased revenues, and that means increased prices for the products and services we sell. But to these unavoidable costs we will also have to add the costs of society's mistakes. Today, the consumer is paying a high price for past policies that almost ruined the domestic coal industry and that have badly crippled the nuclear power and natural gas industries. Tomorrow, we will pay for whatever mistakes we make today.

Let me try to relate the cost factor directly to marketing conditions in New York City. Any motorist who pulls up to a gasoline pump within the city limits, and particularly in Manhattan, will notice that gasoline prices in general are higher than in outlying areas. But he probably doesn't know why.

Again, it's a matter of cost.

My company, for example, has a company owned and operated service station on First Avenue at 37th Street, east of the entrance to the Queens Midtown Tunnel. It's a large, three-bay station on a good sized plot of leased land, and it's kept open round the clock. Now, the cost of locating a similar station at a good rural location on the Interstate Highway system would probably run about $100,000 to $125,000 to buy the land and as much again for the building and facilities. To yield a reasonable 7% or 8% return on that investment, such a station would have to show an after-tax profit of about $1,300 to $1,700 a month.

But the property our First Avenue station is built on is worth more than $1,000,000. The cost of leasing it, plus the higher taxes and operating costs in the city, requires us to charge our customers a few cents more per gallon.

This is the unavoidable price of doing business in Manhattan. If the consumer is unwilling to pay it, he can buy our products outside the city limits, and a great many resident consumers do just that.

But there are also *avoidable* costs of doing business in this city that, unhappily, are *not* being avoided. They make it unnecessarily expensive to live, work, and buy here, and they drive consumers and their purchasing power out of the New York City market.

One avoidable cost, which we may encounter in the not too distant future, is the increased price of gasoline that will result if the recommendations of the State Environmental Department concerning lead additives are accepted. Commissioner Henry L. Diamond is reported as saying that the levels of airborne lead in New York City are regularly "so high as to pose a risk to public health." He is, therefore, recommending that the New York Metropolitan area limit lead content of gasoline sold here after January 1, 1973, to one-half gram a gallon and that all gasoline sold in the area be lead free after January 1, 1974. These regulations far exceed those of the Federal Government.

If the Commissioner's charge could be documented, it would be a serious matter, indeed. But I know of no valid evidence to support his claim, and I feel that before the citizens of New York are saddled with these extra costs, they are entitled to some more adequate proof of such a broad statement.

Now some might ask, "What does it matter? Even if he can't prove a health hazard, let's cut down on lead anyway. What harm would it do?"

Well, as far as the economy of this city is concerned, it can do considerable harm. There are approximately 80 million cars on the road in this country manufactured before 1971 and requiring high-octane gasolines. It will be another 10 years before these cars are completely replaced, and during that time they will require gasolines ranging from 94 to 101 octane. If the owners and operators of these cars in New York City are forced to buy low-lead and no-lead gasoline, they may have to pay up to 10 cents more a gallon for fuel.

For that matter, there are undoubtedly a great many avoidable costs involved in the whole approach that is being taken to prepare to meet automotive emissions standards in 1975 and 1976. To my knowledge, no practical emissions control system has yet been devised that will meet the standards set by Title II of the Clean Air Amendments of 1970, with or without the use of leaded gasolines. Nevertheless, automobile manufacturers are now considering the installation of additional control devices on 1975 and 1976 model cars that would raise new car prices by $600 to $1,000 and would also require the use of higher cost no-lead gasolines. Installation of these devices will also increase gasoline consumption, since they have the side effect of reducing engine efficiency. In my opinion, it would be extremely wasteful to embark on a program of this magnitude before completing a full investigation of other alternatives and an analysis of the costs and benefits of each.

I am sure that the environmentalists would insist that their objectives are as vital to the future welfare of society as is energy. But I seldom hear them discuss costs or relate costs to the value of the benefits they seek. Every environmental improvement we make must be paid for in cash. And every cent we spend on environmental programs, which are not cash generating, must be diverted from other social and economic programs. Environmental quality belongs high up on our list of national priorities, but it's not up there alone. We need to plan well and spend wisely.

We must remember, too, that the rest of the world is watching what is happening here. In the United States we often experience social, economic, and political change in advance of other countries. And in the United States, New York City sets the pace. In some ways, that's very flattering. But it is not always an enviable role. The men and women responsible for charting this city's future often have to solve problems no other city has ever faced. They have to break new ground without the advantage of past experience.

The task calls for imagination, wisdom, and—yes—hard work. It also requires a cooperative effort on the part of all those who have a real interest in social and economic progress and in the well-being of our citizens.

We won't solve the problem by opposing each other. We won't solve it by going our own separate ways. We will have to work it out together. By making a serious effort to understand and appreciate all the many viewpoints involved, we will hopefully achieve that elusive overview that will give us both the acceptable enironment we all want and the adequate supplies of energy we all need.

## Introduction of Milton Musicus

*Clifton Daniel*

I remarked earlier that Gordon Griswold, being a good Brooklynite and Jerseyite, was obviously born somewhere else. Robert Rickles, David Sive and Milton Musicus were all born in Brooklyn.

Mr. Musicus has spent his entire career in public service, and it has been a distinguished career. He is a recipient of the Alfred E. Smith award for outstanding public service in New York State.

He began his government career in the modest post of examiner with the New York City Department of Personnel. He is now administrator of the New York City Municipal Service Administration, and also Director of Construction. But his very special reason for being here is that he is chairman of the Mayor's Interdepartmental Committee on Public Utilities, which is responsible for developing plans for the improvement and maintenance of public utility services in the city.

Between two periods of service with the city, Mr. Musicus was in state government. From 1959 to 1960, he was assistant secretary to the Governor for Administration. During that time, he prepared a plan for the reorganization of the executive branch of the government, and was responsible for the administrative services of the Executive Chamber and the formulation of a number of the Governor's programs.

Aside from his services to the city and state, Mr. Musicus has been a consultant to a number of foreign governments, particularly on problems of governmental reorganization.

Milton Musicus—representing city government.

# ADDRESS

*Milton Musicus*

It is most appropriate that the Board of Trade, in sponsoring a meaningful dialogue on the subject of energy, has recognized its importance to the economic progress and welfare of our City.

For the City to survive, it must have energy to support basic human needs—to provide light at home and work, to move people and goods both horizontally and vertically, to operate machinery to produce goods and services, to pump water and treat sewage, to operate our system of communication, and to do thousands of other jobs which increase human productivity and make life more pleasant.

Our commerce and industry and our quality of life are dependent upon the availability of energy at reasonable costs.

Unfortunately, shortages in fuels, the unfulfilled promises of "clean" and abundant nuclear power, and environmental concerns have left us aimlessly adrift in uncharted waters.

The public is confused by uncertainties, frightened by dire predictions, angered by rising prices, and frustrated by lack of leadership and direction.

Everyone recognizes that the situation is bad, but there still is no national policy on energy. There are no priorities for allocation of available fuels and the uses to which electricity is to be put. Above all, there is no real support for research.

Instead, we are told that if prices are raised, there will be more natural gas and oil discovered in our country. We are also told that limited opeation of nuclear power plants that do not fully meet the requirements of the National Environmental Policy Act would make more electricity available without harmful consequences.

It is a case of the solution being the problem.

We must not permit our anxieties to blind us to the irrationality of simplistic antidotes to complex and involved relationships.

On the other hand, we cannot hide ourselves in the "safe" refuge of lofty idealism. I do not wish to be labeled an alarmist but, in my opinion, we have come to the crossroads at the point of no return.

The City must, as best it can, find its own salvation. And in so doing, it must take into consideration that it is an area of about 600 square miles or about 1/80th of the area of the State of New York, but it is populated by approximately one-half of the population of the State, and consumes about one-half of the energy. This poses special environmental requirements.

The City faces other problems over which it has no control. It can do little, if anything, about curtailment in shipments of natural gas, the rising cost of gas and oil, or the approval of new electric generating plants.

Our only recourse is to promote and influence the use of our energy resources—oil, gas, electricity—for those purposes for which they are essential, and for which they are proper to satisfy environmental standards. We cannot indulge ourselves in the luxury of favoring any one source of energy because of its convenience or lower cost.

Natural gas is essential for the health of the City's millions of inhabitants because it represents the best available means of reducing air pollution. According to the Regional Administrator of the U.S. Environmental Protection Agency, any Federal action which would tend to diminish the already inadequate supply of gas in our City would increase air contamination and result in adverse effects on human health.

To reverse the present trend of curtailing natural gas supplies to the City, there must be not only an increase in supply of gas but a national redistribution of the supplies that are available.

There must also be a revision of the gas price structure which adversely affects the City.

An industrial customer in Brooklyn who uses one million cubic feet of gas per month on a firm purchase agreement pays $1.48 per one thousand cubic feet.

A customer with similar requirements pays $0.80 per thousand cubic feet in Charlotte, North Carolina and $0.46 in Alexander City, Alabama.

These rates make the use of natural gas in the South very

attractive for industrial purposes. It is no small wonder, therefore, that the City of Laurens in North Carolina commits about 70% of its gas puchase to firm industrial customers.

We have, therefore, the ironic situation of a fuel needed to preserve health in one part of the country used for industrial purposes in another because of its low cost.

The supply and demand problem of natural gas is becoming so increasingly critical that the Federal Government must participate in the delicate and difficult process of balancing environmental and economic needs during this period of dwindling supplies.

For our part, the City is fully supporting the importation and local storage of liquid natural gas, and the manufacturing of synthetic natural gas.

To protect air quality, it has mandated a drastic reduction in the sulfur content of oil.

But in the case of oil, as well as natural gas, the City is concerned about rising costs and reliability of supply.

We have, therefore, advocated a change in Federal policies on leasing lands for oil and gas exploration. We have also urged increased research into developing new sources of oil and gas using the nation's vast deposits of shale and coal. We have also pressed for repeal of the oil import quota.

When we turn to electricity, we find that the City is not only fully dependent upon this source of energy already, but that its need for power will increase.

Our efforts to increase job opportunities, encourage new construction, expand our transit system and air condition subway cars, added to the demand for air conditioning in homes and places of work, will increase the need for electricity.

We are so conscious of the large commercial enterprises in the City occupying high-rise office buildings that we frequently overlook the fact that we have many small firms in New York.

These small firms, engaged in a wide range of industries, greatly depend upon electric power for operation. They produce apparel, printing, electric machinery and other products employing women and minority workers.

It is crucial to the City's job development program that such blue-collar industries remain in New York City and expand, because they offer hundreds of thousands of jobs to unskilled workers.

The City is, therefore, engaged in a concerted effort to develop large industrial parks that will contain multi-story industrial buildings and also to help manufacturers expand their existing plants on their present sites.

There are also plans for a number of large developments of sorely needed middle and low-income housing.

In addition, the City's construction program includes 50 miles of new subway lines, 600 additional new air conditioned subway cars and 700 non-air-conditioned cars, sewage treatment plants, schools, courts, colleges.

An inadequate or unreliable supply of power will retard this progress, and is likely to cause a commercial and industrial exodus from the City and discourage new business from locating here.

An examination of Consolidated Edison's power generating system and its plans for the future reveals that the available reserves of electricity should begin to increase in 1973 and continue to increase through 1976. Unfortunately, there are no plans nor approved sites for construction of additional power plants that would start operation in 1977 or thereafter.

What is even more disturbing is the quality of the system Consolidated Eidson is now operating.

Fully 1,646 megawatts of Consolidated Edison's existing generating capacity is produced by 32 machines that are obsolete. Thirty of these are over 40 years old, and, at best, cannot be operated at full capacity for extended periods of time.

Nearly 2,000 megawatts are generated by gas turbine generators and 348 megawatts more of such capacity will be installed in July.

These two categories of plants constitute almost one-half of Consolidated Edison's generating capacity. They are constantly in need of repair, waste fuel, are costly to maintain and operate, and contribute to the City's pollution.

Beginning in 1977, it is expected that Consolidated Edison's reserve capacity will start to decline, and the reliability of its system will probably be poorer than it is today.

The power siting bill recently passed by the State Legislature may expedite somewhat the determination of power plant locations. There will remain, however, the need for vast amounts of capital outlay. It is for this reason that the City has been advocating that the entire State be considered a single system for generating and transmitting electricity, with the Power Authority of the State of New York being the supplier of electricity to the New York Power Pool. The Authority should have the power generating and transmission capability to sell electricity to those utilities which face difficulties in meeting their local needs.

The City has also urged that there be a contract between the Power Authority and the City for purchase of 1,800,000 kilowatts of low cost electricity to meet its governmental needs, to attract industry that can employ and train low-skilled employees, and to serve public housing.

We were, therefore, pleased to learn that in the past month PASNY agreed to sell up to 150,000 kilowatts of power to Consolidated Edison during the summer period to relieve the peak-load demand for electricity.

We were also pleased to learn that the State Legislature has authorized PASNY to construct power plants to supply electricity to the largest single customer of electricity, the Metropolitan Transit Authority. Although this isn't all we asked for, half a loaf is better than none.

To speed up the construction of power plants, the City has urged the Federal Government to develop a one-stop system of approving them at the State level to eliminate current delays and duplication of effort. It is hoped that the experience gained under the new State power siting bill will lead to such a Federally sponsored plan.

In a press release on May 3rd, Mayor Lindsay stated that, "until safe, reliable and more environmentally acceptable sources of energy are developed, the only way we can protect ourselves against the disastrous prospects of fuel shortages, intermittent and perhaps prolonged power brownouts and blackouts is to make the best use of what fuels and electricity we do have."

At the request of the Mayor, the City has, therefore, embarked on the following program:

1. We are reviewing with Consolidated Edison consequences of electric resistance heating. Until such time as nuclear fuel can be used to generate electricity, the process of converting oil to electricity and then transmitting the electricity for the purpose of providing heat is a very wasteful process.

What is required is a whole new concept, the use of fuel on an integrated basis so that waste heat is no longer released to pollute the atmosphere and our rivers but, instead, is saved and used to heat and cool our buildings.

This is now entirely possible because of modern technological developments.

2. We are urging that all new major realty developments in the City, public or private, include provisions for on-site electric generating plants designed so that the waste heat is salvaged to provide heating and air conditioning for the project (see Figure 10). In addition, the possibility of using collected solid waste disposal as a source of fuel and developing potable water is to be considered as part of the system. In comparison with electric resistance heat, this could raise the use of fuels from 30% efficency to 70%.

3. We are instituting a requirement that all equipment and systems requiring fuels or electricity that are included in the design of new municipal buildings or the alteration of existing ones will be judged not only on their initial costs, but also on the basis of the cost of energy required during the life of the installation. City agencies planning new buildings and new equipment will have to present operating cost justifications to the Budget Bureau in addition to first costs.

A start has already been made along this line. New York City recently awarded contracts for the purchase of room air conditioners to vendors because the equipment selected will produce a saving both in money and electricity over the life of the machine.

In some instances, we bought higher priced machines because during the expected life of the air conditioner, the City would save money. In addition, it would help to conserve electricity.

For example, in evaluating the bids received from vendors for air conditioners in the 15,000 to 16,000 BTU group, the lowest bid on the machine rated to produce about 15,000 BTU of cooling was $256 per unit, but it consumes 2,850 watts per hour.

Another vendor offered an air conditioner that produces 16,000 BTU at a price of $292 per unit, but this machine uses only 1,840 watts per hour.

The more expensive machine will cost the City $7.75 more a year for the difference in cost plus interest but will save $30.15 a year in electrical operating cost.

4. We are considering changes in City building codes and other regulations to require that all new construction and renovation provide for windows that can be opened and also meet

Figure 10. Total Energy Plant Built in 1965 for Warbasse Houses, Brooklyn. (Courtesy of Brooklyn Union Gas Company)

energy conservation standards on lighting and insulation.

5. We will develop efficiency standards for all electric appliances, together with a system of labeling appliances in accordance with the standards. This has already been done for air conditioners by the manufacturers (see Appendix).

6. We are planning a broad consumer educational program on energy conservation for distribution to libraries, schools, colleges and other outlets and for broadcast by the Municipal Broadcasting System.

7. We are examing the desirability of authorizing building owners who purchase electricity at bulk rates to install direct metering of their tenants.

At the conclusion of the Mayor's press statement, he pointed out that "we are going to have to make some hard decisions in balancing our social, economic and environmental needs. The scarcity of energy now and for the next several years at least will put increased strain on our ability to satisfy the public demand for energy."

The program presented to you today will not produce an increased supply of fuels and electricity nor will it reduce costs. As a matter of fact, the cost of energy is bound to increase rapidly in the years ahead.

The best we can offer as an approach at this time, is to allocate our existing resources wisely, slow the growth in energy use, and encourage research in technology which would make possible a reconciliation of economic growth and an acceptable environment.

# Introduction of Joseph C. Swidler

*Clifton Daniel*

Joseph C. Swidler assumed his duties as Chairman of the Public Service Commission on February 1, 1970 by appointment of Governor Nelson A. Rockefeller.

Born in Chicago, Ill., on January 28, 1907, he received his early education in schools of that city, including the municipal Crane Junior College.

He subsequently studied at University of Illinois, University of Florida and University of Chicago from which he was graduated with the degrees of Bachelor of Philosophy in 1929 and Doctor of Jurisprudence in 1930.

Upon his admission to the Illinois bar, Mr. Swidler was employed in the Chicago law office of David E. Lilienthal until 1932 when he entered private law practice. In the same period he was also an Associate Editor of the Chicago Commerce Clearing House's Public Utilities and Carriers Service.

In May, 1933, he moved to Washington to serve as an Assistant Solicitor in the Department of the Interior. From that post, later in 1933, he moved to Knoxville, Tennessee, to work for the recently created Tennessee Valley Authority, serving first as its Power Attorney.

He remained a member of the TVA legal staff until 1941 when he returned to Washington on loan to serve as Counsel for the Alien Property Bureau of the Department of Justice and later, also on loan from TVA, as Counsel for the Office of War Utilities of the War Production Board.

In the period 1943-1945 Mr. Swidler served in the Navy, first as a seaman in the Seabees. He was later commissioned, and served as a Lieutenant on the staff of the Assistant Secretary of the Navy in Washington.

Upon his return to civilian status, Mr. Swidler rejoined TVA as its General Counsel and Secretary, serving also as Chairman of the Board of the TVA Retirement System, continuing in those posts until 1957 when he left TVA and resumed private law practice, first in Knoxville and later in Nashville, specializing in utility matters.

In 1961 he was appointed by President John F. Kennedy as a member and Chairman of the Federal Power Commission, continuing in that office until December 31, 1965 after which he established a law practice in Washington, D. C., in which he was engaged until he assumed his present position of Chairman of the Public Service Commission. In 1964-65 he was also a member of the U.S. Water Resources Council.

In 1944 Mr. Swidler was married to the former Gertrude Tyrna. They have two children, Ann and Mark, and now reside in Loudonville, a suburb of Albany.

# ADDRESS

*Joseph C. Swidler*

Energy supply and its environmental impact is one of this country's most difficult and controversial domestic problems. Almost nothing that can be said about it is cheerful or comfortable. Energy forums, conferences, seminars, workshops and colloquiums are being held over the country on virtually a nonstop basis. This meeting gains distinction only because New York City's energy problem is more acute than that of the rest of the country, but the problem and the concern are nationwide.

While the interest in energy problems is intense and almost obsessive on the part of many groups, most people are only dimly aware of the profound changes which are taking place in energy supply. Energy is still fairly cheap. The consumer has not yet suffered severely from the impact of shortages. Naturally enough, he wants to keep things as they are, and resents talk of price increases, or interruptions of energy supply, or the threat that some utility facility might be located in his parish or line of vision. Public figures are not anxious to disturb him.

The energy problem of this country is particularly baffling because it is so new. It seems to many to be somehow un-American and subversive to talk of energy shortages in the country that has boasted through the last half century or more that it had the highest energy use in the world and a corresponding degree of general affluence. With about 6% of the world's population, this country accounts for almost 40% of world energy consumption. But in less than a decade, in an instant of history, this nation has been transformed from one of apparent energy surpluses to one of large and growing energy deficits.

The change in our energy posture is made graphic by consideration of the changes in the ratios of reserves to production from natural gas, which supplies a third of total national energy needs. In 1950 the known reserves of gas were 185 trillion cubic feet (TCF) and the annual production of gas was 6.9 TCF, so that reserves were almost 27 times annual use. In 1971, annual use was over 22 TCF. Excluding Alaska, reserves were 247 TCF, and the ratio of reserves to production dropped to 11.

The gas shortage, in the Northeast in particular, is not a matter of exhausting reserves at some future date, but of facing now—in the coming winter, and in the winter of 1973-1974—the inability of the pipelines to bring in enough gas to satisfy the demands of industries, commercial establishments, apartment houses, and homeowners. No new supplies of gas have been made available in the New York market since 1970.

The gas companies are attempting to augment supplies by purchase of liquefied natural gas (LNG) from abroad and by building synthetic gas plants in this country, but the cost is two or three times the present price of natural gas from domestic underground sources and the expansion of supply entails an extended lead time. Both LNG and synthetic or substitute natural gas (SNG) involve dependence on foreign sources. In the 1980's we shall probably see the advent of pipeline gas derived from domestic coal sources, but at high cost and too late to help in the near term.

The deterioration in this country's oil situation has been masked by steady increases in imports. It may surprise you to know that more than a quarter of this country's oil supplies already are imported. A few days ago the White House revealed another increase of 230,000 barrels a day in oil import levels. There will be more such announcements, because increasingly large oil imports are no longer an option but a necessity. Of course, oil imports for generating electricity on the East Coast will not increase substantially in percentage. That is because the East Coast is already importing 93% of its boiler fuel needs.

The National Petroleum Council of the oil industry forecasts oil consumption in 1980 of 22½ million barrels a day, of which about 12 MBD would represent domestic production and some 10.7 MBD, or 48% of total oil requirements, would be imported. This is serious enough, but the New York State Public Service Commission staff thinks that the National Petroleum Council has overstated the amount of energy which will be derived from nuclear power and coal by 1980 and that the shortfall must be made up by oil. For example, the Council forecasts an expansion of present coal use by a third. We think it will decline. Our estimate is that oil requirements by 1980 will be 28.3 MBD, and that imports will probably reach 16½ MBD, or 58% of 1980 national oil requirements and approximately equal to the present total national usage. The implications of this degree of dependence on imports are unsettling, especially considering that most of the shortfall must come from the Middle East and North Africa. We estimate that by 1980 11½ MBD of oil imports will come from this troubled area of the world. One hardly dares think of the impact on our economy and national life if so large a share of national fuel requirements should be cut off in one of the perennial Middle East political conflagrations, because no longer do we have spare capacity in domestic producing fields, in refinery output, or in transportation facilities, to make good the shortages for more than a few days.

The risk of interruption of imported supplies is not the only reason to fear for adequacy of oil deliveries. We may run out of money. By 1980, the annual cost of fuel imports, even at current prices, is likely to be more than $13 billion.

The story with respect to electric generating capacity is different. Electricity is not a limited and exhaustible resource like gas and oil. Putting to one side the question of uncertainty of fuel supply, adequacy of electricity depends on the construction of enough generating units and transmission lines. The serious problems we are having with siting of plants and transmission lines warn us that critical shortages of electric service are in prospect for the whole country. I fear that New York City will be among the first to face this crisis.

The nation's electric utilities reached an installed generating capacity of 340,000 megawatts in 1970. By 1980 we shall need to have added about 325,000 MW of new capacity, bringing the total to 665,000 MW, or a near-doubling of capacity in just ten years. Because the time needed to design, construct and license a nuclear plant has risen to about eight years, and seems likely to go to ten years unless the licensing process can be improved, it is obvious that some 350 or more major new units, mostly nuclear, should be in construction or nearing construction right now, but less than half are yet under way. I say mostly nuclear because there is not enough gas or dependable supplies of low-sulfur oil to run all of this new capacity, and coal is outlawed in much of the country.

It takes almost twice as long now to build a power plant in this country as in the rest of the world. The delay results in extravagantly high costs, which are reflected in rates. This is not a cause for pride in the American capacity for getting things done.

New York State has about 6% of the nation's total generating capacity, or about 23,000 MW. We shall need to add about 20,000 MW to our capacity in this decade. I am not optimistic about our ability to do so. In the period from 1967 to 1971 the State's utilities were able to bring on line only about 3,400 MW of major-unit capacity, although some gas turbines were also added. Beginning or completion of construction of a number of large new units, aggregating about 5,400 MW, is now delayed. Unless we can overcome the design and construction problems and the licensing delays that beset our new generating units, we shall not be able to serve the electric loads of the State projected for 1980.

In New York City the prospect is even grimmer. The installed capacity of Consolidated Edison Company is about 8,800 MW, much of it antiquated and unreliable. For the summer of 1972 the company's load is predicted to reach 8,400 MW, leaving practically no reserve margin for outages of machines. Such outages last summer averaged about 1,400 to 1,600 MW and on occasion rose to about 2,400 MW. The company hopes to bring about 350 MW of new gas turbines on line by mid-summer and to purchase 700 to 800 MW from other utilities, but the reserve would still be too low to provide assurance of continuous supply. It is possible that the new unit at Bowline Point on the Hudson, a 600 MW plant owned jointly with Orange and Rockland Utilities, may come on line in July or August, but new units require a breaking-in period of several months before they can be considered reliable.

For the longer term, Consolidated Edison will need to add about 5,500 MW of new capacity by 1980 while retiring about 1,900 MW of old units, for a net addition of 3,600 MW. Of the 5,500 MW of new requirements, about 3,600 MW has been started or is covered by AEC license applications. Whether it will succeed is problematical. Some of this new capacity has not yet been sited and very little of it has the necessary permits or licenses. Only a determined optimist would predict meeting this objective in the face of compounded difficulties.

The picture for electric power supply in New York City is

stark. The time has come when it is no longer possible to take for granted that expansion in power loads may continue and that somehow or other the loads will be served. I see announcements from time to time of great new programs for urban development within New York City, and I wonder whether the people who are proposing these badly needed programs have given enough thought to the question whether they will be able to obtain an adequate and dependable energy supply.

Basically the problem of shortage in electric generating capacity is a result of concern about the effects of electric power generation on the environment. Only a decade ago it was a relatively simple matter for a power company to build a generating plant. Licensing requirements were relatively few, and there was little opposition. The plants have grown in size and number, with increasing impact on the environment. Growing population has created new siting problems, at the same time that environmental standards have become progressively more severe.

At first, public opposition focused on the air pollution problem, and it seemed as though nuclear energy, which does not contribute to air pollution, had come to the rescue just in time. The power industry in great relief turned to this energy source. Unfortunately, it suffers from the same problem of thermal effect on surface waters as the fossil fuel plants, and in somewhat greater degree. Moreover, it introduced a new environmental complication from the hazards of radiological releases. Nuclear stations require licenses from the Atomic Energy Commission under a complex of Federal laws, and AEC's mills grind slowly. In fact, they have almost ground to a halt. AEC has some 40 construction licenses under consideration but has issued no new construction license in the last 13 months. The siting of nuclear stations has, therefore, proven to be even more difficult than that of conventional power plants.

It is understandable that warnings of an energy crisis have created no general alarm. There is as yet little visible evidence of the seriousness of the problem. It is also understandable that when most laymen do face this unappetizing picture, they would either seek a scapegoat or probe for some easy solution. An irate citizen will tend to blame the whole problem on the people who use electric toothbrushes and drink beer from aluminum cans, and solve it by proposing instant construction of solar energy systems, preferably in outer space.

The indispensable first step toward dealing realistically with the energy crisis is to reach a consensus on a few fundamental propositions. I believe most informed citizens could agree that the nation's petroleum reserves are being depleted below safe levels; that there is no presently available solution which will make electric power cheap, plentiful, and free from any adverse environmental impact or drain on nonrenewable resources; and that there is no way to suppress all load growth without intolerable damage to our society. We need to move quickly to expand our domestic energy sources and to build new generating capacity, at the same time that we make more effective use of energy supplies, and move vigorously to improve energy technology, both for conservation and environmental purposes. Environmental standards should be high but not arbitrary, and common sense is just as necessary on environmental matters as in the other affairs of life.

Many people shrink from a comprehensive and resolute program which involves construction of power plants and transmission lines, and suggest purported solutions which would make it unnecessary. Let me mention some of them. One is somehow to do away with the appliances they label "frivolous," such as the electric toothbrush, hair-curler, razor, and can opener. Unfortunately, all have an infinitesmal energy use, in the order of a thousandth of average annual residential use.

The proposal frequently heard for an inverted rate structure, one where unit rates increase with usage, suffers from the same fault. The Public Service Commission studies show that in the short term at least no conceivable increase in rates would be likely to depress residential demands significantly.

Another frequent proposal to prevent the growth of power loads is to include in the price the full environmental and social costs of mining coal, drilling for oil and gas, and generating electricity—"internalizing" these costs, in the current jargon. Aside from doubts as to the efficacy of such cost increases for the intended purpose, I know of no way for state or federal utility regulation to impose these social costs until they are reflected in actual expenditures by the utilities. As the costs of protecting workers and the environment are more fully imposed on fuel producers and electric utilities, fuel prices will rise and so will the rates. The process of internalizing environmental costs is well under way, and has been reflected in recent rate increases. More will follow. I might add that thus far consumers do not seem pleased with the rate effect of the internalization process.

The Commission receives thousands of complaints about rate increases, but we have yet to receive a letter of commendation for the fact that environmental costs are now a substantial factor in rising power prices.

While I do not favor inverted rates, I do favor the elimination of promotional rates, that is, rates where the quantity discount is greater than can be justified on a cost basis.

Finally, there is the evasion based on insisting that if a power plant is to be built it must be of a presently impossible kind, such as a solar power or fusion plant. Solar energy is a particular favorite. I do not intend to disparage this ultimate energy source, but you can gain a measure of the practical problems in the way of its realization by considering that a presently preferred approach is to use giant satellites to receive the energy from the sun and electro-magnetic beams to transmit the energy to receiving and transforming stations on earth. Installation (if that is the word), repair and maintenance would be achieved with space shuttles and space tugs. Mind you, these space installations and vehicles must be scaled to vast dimensions if they are to be more than scientific curiosities. Solar energy systems may come to pass, but not without the commitment of many billions of dollars and the passage of many years. Relying on the best information available to me, it appears that solar power is decades away at best. Fusion power is closer to assurance of realization in time, but 1990 would be an optimistic target.

As of here and now, the practical choices are limited. We may have fossil fuel stations, fired with oil or coal, or nuclear plants. Pump storage plants also are needed, but they merely serve in effect to store energy generated with fossil fuel or nuclear power, and do not replace these sources. Conventional hydro sites are too scarce in the East to be a factor, and present environmental problems of their own. Imports from Canada are a possibility, but only for a small share of needs. After examining all other real possibilities, we are back to the need to build fossil fuel or nuclear power plants for major load requirements. The environmental choice is between the products of combustion and the products of radiation, minimized in both cases as far as feasible. For the short term, there is no other practical choice.

Many earnest people advocate putting a ceiling on electric load growth and perhaps on all energy growth, rather than permit power expansion. Their view is that if no more power or energy were made available, somehow (they are vague on details) our society would accommodate. This view springs more from hope than reason. There is a close correlation between variations in energy use and variations in gross national product, that is to say, overall output, and the leveling off of energy demands can only be purchased at the risk of economic catastrophe.

The economists for the Public Service Commission have attempted to determine realistically what reduction in the rate of energy growth could be achieved by 1980, assuming that we dedicate ourselves to the effort but avoid measures which would significantly reduce the standard of living or cause widespread unemployment. Their conclusion is that by improving the gasoline mileage of passenger automobiles, greater efficiency in power generation, and greater economy in the climate conditioning of new buildings, the rate of growth could be reduced by about an eighth, from 4.2% compounded to about 3.7%. This is by no means a negligible reduction, and it would make possible a 7% saving in overall energy use by 1985.

Reduction in the rate of growth by an eighth would not come effortlessly. For example, to achieve the transportation savings would mean increasing the average miles per gallon on passenger cars from the 1969 average of 13.75 miles per gallon to 20 miles per gallon, necessitating a radically different mix of automobiles than we have today. The average efficiency of electric generation in fossil fuel plants would need to be increased from 35% to 37%. The energy requirements of new high rise buildings are assumed to be only half that of present structures, because of superior insulation and other energy conservation features.

Further along in time it may be possible to loosen the tie between an improved standard of living and large energy growth, or we may develop energy systems which relieve our concern about the impact of energy growth, but within the planning period of the next two decades environmental compromises are inevitable if our society is to remain viable.

I have probably sufficiently indicated what I thought should not be done under the guise of remedying the energy crisis. Before concluding, I should mention some of the things which should be done. Completion without delay of 3600 MW in generating projects already under way is the paramount need. Municipal support and the support of labor, consumer and business groups may prove essential. Their voice is rarely heard in permit and licensing proceedings.

A few weeks ago I would have placed near the head of the list the enactment of a state power plant siting bill, although Federal siting legislation is even more important. On the first of the month the legislature passed such a bill, it has been signed by Governor Rockefeller, and it takes effect on July 1st. Badly as this measure is needed in order to establish a workable procedure for siting new power plants which are subject to state jurisdiction on the basis of a reconciliation of the need for expanded power supplies and for the protection of the environment, its benefits will not appear until late in this decade.

Improvement in the administration of Federal licensing administration is a key need, but the subject is too complex for adequate treatment on this occasion.

The adoption of a comprehensive energy conservation program stands high in priority. Reducing requirements by the elimination of waste is a good deal better than building capacity and burning fuel needlessly.

Governor Rockefeller has recently created a State Interdepartmental Fuel and Energy Committee, consisting of the heads of the Departments of Public Service, Environmental Conservation, Commerce, Transportation and the Offices of General Services and Local Government, to organize a State program of energy conservation. It is my hope that before the Legislature meets again this Committee will be able to present to the Governor a comprehensive program of statewide energy conservation measures applicable in both the public and private sectors, and that this will be followed by requests for appropriate legislation. The energy problem is a national one, and it cannot be solved without a coherent and comprehensive Federal approach to the problems of energy supply and demand. Nevertheless, there is much that can be done at both the State and City levels.

This country needs a greatly expanded research program with emphasis on such short term goals as coal gasification and liquefaction and stack emission controls in order to make our vast coal supplies usable without violence to air pollution goals; shale oil recovery; the breeder reactor; combined cycle gas turbines, superconducting generators, fuel cells and other ways to improve the efficiency of generation; better ways to dissipate the waste heat produced in the process of power generation; and more economic methods of underground transmission, including superconducting cables. For the longer term, we must seek energy systems, such as fusion and solar energy, which hold promise of relieving the pressure on fossil fuel resources, and of avoiding the choice between the environmental side-effects of combustion or radiation.

I should like to conclude with an unambiguous summary of New York City's power supply situation. The Consolidated Edison system does not have enough capacity to serve existing loads, and the capacity under construction is less than is needed to serve future growth. A number of construction projects are bogged down, mostly on account of permit and license problems. Power conservation measures are essential but they will not bridge the gap. Futher delays in the schedules for completing units under construction could result in disaster. Beyond that, two major units or the equivalent should be scheduled now, one for completion in 1978, and the other for completion in 1979 or 1980. Without evidence of progress on projects under way and early starts on this new capacity, it will be necessary to restrict loads to the level which can be served with whatever generating units are available.

# Workshop No. 1
# CRISIS IN ENERGY SOURCES AND PRODUCTION

*Moderator*
John Noble Wilford
*The New York Times*

*Federal Government*
Stephen Wakefield, Deputy Assistant Secretary
Energy Programs, Bureau of Mines
Department of the Interior

*Environmental Protection*
Joseph Kearney
Natural Resources Defense Council

*Electrical Industry*
Frederick W. Sullivan, Vice President
Consolidated Edison Company of New York, Inc.

*Gas Industry*
Charles Neumeyer, Sr. Vice President
Brooklyn Union Gas Company

*Oil Industry*
Robert Sampson, Coordinator of Natural Resources
Cities Service Oil Co.

## Workshop No. 1
## CRISIS IN ENERGY SOURCES AND PRODUCTION

I. Overview (1972-85)

   1. Primary reliance on conventional fuel sources, problems in research, technology, and lead time for breeder reactors, fusion, solar energy, and coal gasification will delay major contributions from these potential energy sources until after mid-1980's.

II. Oil

   A. NEPA and Other Environmental Issues

      1. Accelerated Government Leasing Programs on Outer Continental Shelf

   B.. Oil Imports

      1. Costs and Risks: economic, national security and environmental

   C. Improved Exploration Incentives

      1. Product prices/foreign imports
      2. Tax incentives

III. Gas

   A. Accelerated Government Leasing of Outer Continental Shelf

   B. Amendments to Natural Gas Act

      1. Security of contract terms
      2. Competitive pricing

   C. LNG/SNG—Imported and Domestic

   D. Interstate/Intrastate Marketing Policy

IV. Electricity

   A. Generating Plant Licensing and Regulations

      1. Nuclear technology
      2. Pumped storage

   B. Availability of Low Sulfur Fuel Oil/Gas

      1. Efficiency of conventional sources

   C. Purchasing Power (Imports)

*MODERATOR WILFORD:* Let us try to confine our remarks and discussion to the near future which we will define as between now and 1985. We have representatives from the government, from the environmentalists, and from the various energy industries and we will start off with Stephen Wakefield, Deputy Assistant Secreatry of Energy Programs, the Department of the Interior. We hope to get from him an overview of the various alternatives for energy sources and production between now and 1985—those that are considered feasible and those that are considered not feasible by the government.

Then we hope to get the environmentalists' input from Joseph Kearney of the Natural Resources Defense Council and then we will go to Frederick Sullivan, Vice President of Consolidated Edison, Charles Neumeyer, Vice President of Brooklyn Union Gas Company and Robert Sampson, Coordinator of Natural Resources, Cities Service Oil Company. We will try to be as informal as possible at the table and as soon as we've made presentations, we'll be open for questions from the floor to the panelists. O.K., Mr. Wakefield.

*MR. STEPHEN WAKEFIELD:* I think we've had laid out pretty well so far today what the problems are and I would like to indicate to you what the thinking of the Department of the Interior is on what the solutions may be, and what within our time frame they almost definitely are not.

First with respect to coal. Coal is by far our most abundant fossil fuel resource. We've got about 600 years known reserves based on last year's production, but coal is being forced out of metropolitan markets by the necessary air pollution requirements. We have a very good and intensifying accelerating program underway both in the Office of Coal Research and the Bureau of Mines, for both the gasification and liquefaction of coal and for other methods of cleaning it up. This is in cooperation with the American Gas Association for high BTU pipeline quality gas. We're getting ready to start a program for the low BTU gas which will be used by industry and in power plants. We are hopeful that this technology will be commercially available about 1981, so within the period of from now to 1985 we can't really look for a significant contribution here.

We have very substantial oil shale reserves that can be made into synthetic oils and possibly gas in the western part of the United States. We also have tar sands, although they're more prevalent in Canada. Again it will probably be 1985 or later before significant contributions are made there. It's a question of both technology and economics in respect to this. As for liquid metal, fast breeder, nuclear reactors, it will probably be 1985 or 1990 before they make a significant commercial contribution. Geothermal and tidal energy, while maybe not so far away, are limited possibilities both as to amounts and to locations. There are only certain locations of these energy sources. Solar energy and fusion are very long term solutions, and I wouldn't count on anything in this century as far as commercial application is concerned.

So we really get down to two choices, I think, over the next ten years or so. One is to import more oil. Our projections at Interior are, that if present trends continue, by 1985 we'll be importing 57 percent of our oil. I think a number of the security and balance of payment problems involved in this solution have already been mentioned today. Let me just add one thing further. There is no assurance that we're going to be able to get that oil if we want it even if we assume stability in the mideast, because our projections are that by 1985 the United States will be requesting the mideast and northern and western African countries to export to us as much oil as they presently export to the entire world.

One further thing I'd say about importing oil—it doesn't give us very much gas. At best we can import a little liquefied natural gas with the oil at about four times the present well head costs in Southern Louisiana. So that really leaves us with our domestic oil and gas reserves. The U.S. Geologic Survey at the Department of Interior estimates that in undiscovered resources for which we have the technology today to define and develop in the United States, there is enough natural gas to supply us for almost 100 years at the present rate of consumption. Now to refine that gas and make it available to the consumer is the problem, and I think we'll see that two things need to be done.

First, economic incentives must be offered. This translates into higher prices for the consumer, particularly for natural gas. Secondly, no matter how much we allow the industry to receive for their oil or gas, they still have to have a place to look for it. Over 50 percent of what we consider the undiscovered resources are under public lands, and that gets back mostly to the outer continental shelves around our country. These areas may be very attractive. We don't know yet, because there hasn't been any exploration there and you never know what you will get until you sink the drill down into the ground. As all of you know, there are many objections being raised by people in the northeast to drilling in the Atlantic outer continental shelf. I have sat in on a number of meetings with state representatives and environmentalists and I've perceived four main objections that are being raised. I'd like to mention just briefly each of these and our response to them. If anybody has any more objections, I'd certainly like to hear them.

The first objection I'd call "general aesthetics." Nobody wants to be sitting at their summer home out on Cape Cod and looking at an oil derrick right off in front of them. Our answer to this objection is that areas from a geological standpoint which are considered attractive are all at least 30 miles offshore, and the really attractive ones are probably 75 to 150 miles offshore. In other words the curvature of the earth would prevent anyone on shore from seeing oil wells or the drilling that's going on.

Second are fears of a disaster such as we had in the Santa Barbara Channel. I can't give you any assurance that there wouldn't ever be another disaster like that. I would point out that in the some 38 years that industry has been drilling in the offshore areas around the United States over 16,000 wells have been drilled. There have only been ten instances that even presented an environmental threat and only three or four of those actually resulted in any environmental damage. Since the Santa Barbara incident, the U.S. Geological Survey has tightened up its regulations and its inspection procedures and is doing considerable study in the area of preventing future incidents. It is also working with the Coast Guard and industry on how to clean up spills should they occur.

A third objection that I've heard is that offshore drilling contributes to pollution in the oceans. The studies that I've seen done by the Coast Guard indicate that the amount of oil in the ocean that comes from offshore drilling is about 2 percent of the total oil in the ocean. On the other hand oil that is in the ocean from tankers amounts to about 29 percent, and since the alternative to the offshore drilling is most likely further imports from tankers, I'd submit that you're going to probably get more oil in the water by going that route than by drilling in the OCS area.

The fourth objection is that drilling might interfere with fishing, particularly in the Georges Bank. On that score I would point out what the experience of the fishing industry in the Gulf of Mexico has been since offshore drilling commenced there about 20 years ago. Twenty years ago there was no commercial, no sport fishing in the Gulf of Mexico. Today it probably has the best sport fishing of anywhere in the country, if not in the world. As far as commercial catches are concerned, from 1960 to 1970 commercial fishing catches in the Gulf of Mexico went up by a third, while off the Atlantic Coast they were off by two thirds in that period of time. The reason behind this is that the rigs form man-made barriers out there that can support algae, plankton and other small animal life which the small fish feed on. The middle sized fish feed on the small fish and the larger fish feed on the middle sized fish. I think that anybody who checks with the fishing industry in the Gulf of Mexico, and particularly the shrimp and oyster fishers, is going to find that they are just not opposed to the offshore drilling at all, but probably favorable to it. I think that this is one area that needs to be looked into. As I say we haven't drilled out there, we don't know what we are going to find. There has been drilling off Nova Scotia which has been very successful, but for right now the Geological Survey is conducting some preliminary studies, bottom sampling, not any drilling in itself, but it is trying to determine more about the geology of the area. There won't be a lease sale held out there until the National Environmental Policy Act is complied with, and until there is a resolution of the pending litigation between the Atlantic States and the Federal Government over who owns how much out there, but it's something I think should be considered, and I think people should look at the facts of the situation. I believe that's all I have for right now.

*MODERATOR WILFORD:* I have one question. Assuming that you have good prospects out in the Atlantic and whatever the regulations are going to be, how long would it take from the time you start exploration, and assuming that there is something there, until the time you could have a pretty large scale operation offshore?

*MR. WAKEFIELD:* Well, it varies, but generally I'd say from the time a lease sale was held, and assuming that the industry could progress pretty rapidly from that point, probably two to three years before first discoveries, assuming there are discoveries made, and probably five to seven years before there was substantial production, since it is virgin unexplored land.

*MODERATOR WILFORD:* Thank you.

*MR. CHARLES NEUMEYER:* You stole a little bit of what I was going to say, Steve, with that last remark. As far as the gas industry is concerned, as you all know we are pushing for the development of the outer continental shelf lands and pushing for the leasing. Of course you heard quite a bit this morning about the long term research that is going on for the other sources of energy that may be developed, such as coal gasification, oil shale and what have you. We feel that if immediate action is taken with respect to the offshore and the continental shelf lands, the schedule that Steve mentioned may perhaps be accelerated. While we may be optimistic, we think that there is an opportunity that some of that gas may be flowing from the offshore lands in, we would hope, from five to six years.

Of course, as you know there has been quite a bit of seismological work going on offshore already and there has been quite a bit of information developed with respect to the Atlantic continental shelf. Of course, as far as the Gulf Coast is concerned, they have been drilling down there for some time, and the structures in the Gulf Coast are well known. If there were an acceleration of leasing in the Gulf Coast that would have a very beneficial short term effect on the gas supply.

*MR. WAKEFIELD:* Let me just say one further thing on the development; five to seven years is pretty much based on historical evidence and with the strong industry push because of the shortages, it could be shortened.

*MODERATOR WILFORD:* O.K. Now for the environmental aspects, Joe Kearney.

*MR. JOSEPH KEARNEY:* Thank you very much. I want to thank everybody who is responsible for my being here this afternoon. I appreciate the opportunity to talk with all of you and give you some of my reflections and those of the National Resources Defense Council on the environmental impacts which are felt in supplying the energy to the New York City area. I would like to give certain comments on a lot of things that have been said, but I cannot say everything in the course of the five or ten minutes allotted. I want to first make very clear that we are discussing here the necessities for the particular alternatives for producing energy. I don't want to minimize the requirements for meeting energy demands. Let's accept the particular demands we must meet leaving the question as to what those demands ought to be to another panel. Let us simply question how we are going to meet these demands. We are going to touch on three particular areas.

The first area is that in which I have the most experience and that is in electricity production. I am on this panel as the representative for natural resources for the Defense Council, however I am a graduate student of M.I.T. and am taking my Ph.D. in nuclear engineering. With regard to electricity production the need as I see it in the near future is to increase the efficiencies of the entire system—getting the fuel, burning the fuel and utilizing the fuel.

Let us consider the question of nuclear power. Joe Swidler has pointed out how much demand nuclear energy will probably have to fulfill in the near future. The planners have been very lax in meeting their own projections to date. However, once we see a resolution of the safety questions which are now being discussed in Washington, and once we see the manufacturers of the nuclear power plants having a part of that decision-making process, then I think we can get to utilizing nuclear power more rapidly. I think the utility should also begin to look not only to the water reactors but also to the gas reactors because of economics. Because of their thermal efficiencies which are higher and therefore put less of a load on the environment, and because of the safety aspects, I think that gas reactors deserve another look.

With respect to New York City, the efficiency of city power can be improved significantly by reviewing again the ties Con Edison has with other utilities. I will reflect on one particular aspect of this and that is the summer to winter peak. In New York City we have a summer peak which is approximately 1,000 or 500 megawatts above the winter peak. In New England, however, the situation is exactly reversed. There is an approximately 1200 megawatt winter peak and therefore good ties with the New England region could effectively reduce the amount of peaking power used by Con Edison. We do think that peaking power increases the environmental impact of producing electricity significantly because all the "peakers" are very environmentally polluting gas turbo plants on the barges outside of Brooklyn. There is only one tie as far as I know between the New York system and the New England system, and the possibility of further ties should be examined.

The second area I wish to get into is the political one. I think we really should feed some input into our legislatures and into our representatives in Washington because many of the decisions on national policy which affect the New York area significantly are made in Washington, and we should have some say in them. I refer to the Jackson committee hearings on national energy policy. How many of our legislators have taken a stand, have participated, have even thought about national energy policy and its impact on New York's energy?

Another problem is in the power plant siting area. There are two bureaus now in Washington on power plant siting and, as Joe Swidler said, any overall effect on the system in the area of power plant siting will come from Washington.

Another point with respect to the Washington resolution of the energy problem is that of research and development. The funding for the energy possibilities in the future is going to dictate what energy sources we are going to be using in the late 1980's and subsequent to that and we ought to be sharing in those decisions.

The third area I would like to comment on is the use of gas and oil. It is obvious that gas is a desirable fuel, environmentally speaking. The question is, of course, where we are going to get it. A number of people here have mentioned the possibilities of drilling on the outer continental shelf. I feel that before we explore this site, we should be sure that we have exhausted our natural gas resources in the continental United States. Before we find out whether or not this is true, we have to demand the lifting of the Federal Power Commission regulations on the price ceiling of gas to give the gas companies more incentive to search for and develop alternate sources of natural gas.

With respect to oil again, it's a matter of evaluating the alternatives and developing the reasons why we should be considering drilling the outer continental shelf. Basically we have been presented with a lot of economic reasons to date for the need for going out there. It is a shortage of supply. When we consider shortages of oil, we come immediately to the question of import quotas. A quick calculation on the basis of the President's Cabinet Task Force on import quotas suggests that when the maximum quotas exist over a period of time, the price for imported oil would decrease to about $2.00 a barrel. Taking a conservative estimate of $2.50 a barrel, the difference is two hundred million dollars a year to New York City.

Thank you very much.

*MODERATOR WILFORD:* Now we'll go to Con Edison and hear something about the electrical industry.

*MR. FREDERICK SULLIVAN:* Well I guess you're going to hear a rather prejudiced viewpoint, because I'm basically Con Ed's gas man, but to the best of my ability I'll tell you about some of our electrical problems.

Now as you are probably quite aware, we are in the business of selling three forms of energy: electricity, gas and steam. To produce those forms of energy right now we buy gas which we distribute directly. We buy oil; we used to buy coal—we burned our last coal in February of 1972. Coal is temporarily, and I like to use the word temporarily, out of our fuel bag. We expect it to come back again in the form of pipeline quality gas made from coal, and possibly low sulfur oil made from coal, and possibly even low BTU gas to fire electric generators of the combined cycle type. Naturally we use nuclear fuel, and we purchase some of our energy. We are not buyers of electrical energy.

Now for the future, where do we see our energy coming from? Well, it's the same four sources: coal, oil, nuclear fuel and purchased gas. But there is going to be a different mix in those fuels. Oil will be low sulfur oil and the light oils for turbine fuel. The gas is going to change radically from our traditional pipeline gas to what we can best term the exotic forms of gases made from coal and petroleum products. I won't infringe on Charlie Neumeyer's territory, and I suspect that he would like to talk about the sources of those gases.

Now we certainly have plans to meet our electrical load requirements. To begin with, we have to look at what these requirements are going to be. We must make a projection. For this summer we estimate a peak load of 8400 megawatts. By the time we get to 1990, if you will allow me to refer our future plans, we expect a load of 17,350 megawatts. Now that's a doubling of the load that we would have the responsibility for and I'll stress the word responsibility. It will be our responsibility to meet that load requirement if growth continues the way we expect it to. Now, are we promoting that growth? The answer is definitely no. We are promoting conservation of energy. We have a Save-a-Watt program going. We stopped promoting the use of all of our fuels probably a year before the rest of the industry, but we still have to be prepared to meet the growth that comes along.

Are we building plants? We most definitely are building plants, and when we look ahead from now until 1977, if we are fortunate in our licensing procedures the electrical energy is going to come from nuclear power. We predict that we won't burn any more oil in 1977 than we burn this year. We won't burn any coal, we'll burn less gas and we'll utilize quite a bit more nuclear fuel in the units that we currently have under construction. One is operating; it has been operating for a number of years. Another is ready to go on line in the summer; it's being delayed by licensing problems, and a third is under construction.

Now we have also installed a great deal of gas turbine capacity. By this summer we'll have 2332 megawatts of gas turbines on line. Now these are in a sense peaking units and of course in our long range planning we are trying to phase out the old fossil fuel units, the polluters that Joe talked about, and the replacements are these gas turbine units. We're having to run these turbines more hours than we would like to now, because of the shortage of base load power, but they are peaking units that are in city sources of peaking power in the future.

We are putting in ties to other electrical systems. We've got about three ties in the works now, one of which is completed and the others are well along so we are importing a lot more energy. The only way we are going to get by this summer is to import more energy than we planned to import, but thank goodness we were able to line it up and have it available to serve our requirements. So the ties are being strengthened; there will be an exchange of power. More and more there will be an exchange of power between not only New York and New England but all the way down into Maryland. There is the PJM tie that has now been activated. It's Pennsylvania, New Jersey and Maryland, so we tie all the way down to those utilities. We tie all the way up into Canada. We tie into New England. We tie into western New York State.

Steam capacity is one of our very pressing problems. It is probably the least well known form of energy that we supply, but we are the biggest steam supply company in the United States by far. The hard thing about steam is that you can't generate it outside of the city, as everyone would like us to do with electric power. It has to be generated near where the load is because you just can't send steam a long distance. We have the problem that many of our steam plants are very old. They have got to be replaced. We are trying to replace them. We're up against a situation right now where we are building a plant that we may not be allowed to operate and if we don't operate it we will not be able to meet the steam requirements of New York City this coming winter. So that is that particular situation.

Now gas-wise we were in reasonable shape on our gas supply. We're subject to curtailments like everybody else. We are negotiating for imported liquefied natural gas. We are after a synthetic natural gas supply from a neighboring company. We are supporting coal gasification research and development because we are certain that our long term solution to the gas supply problem is coal gasification. We are not promoting gas, of course, and our load growth has dropped off sharply because of the lack of promotion and the supply restrictions enacted and applied by the Public Service Commission. So it means that our gas will go a little further towards meeting the needs of our customers. Of course this existing gas is going to take a widely different form because it's no longer going to be traditional gas, it's going to be a mixture of traditional gas and synthetic gases. It's going to get very expensive. We see the gas from our supplier going up to about a dollar by 1980. As against that our average cost of gas this year was about 45 cents per million BTU's.

We're paying 70 cents a million BTU for the lowest sulfur oil we're having to convert to now. Of course, the same sulfur oil affects our steam generating capability, and it ends up in increasing the cost of the service to the customer. Well, I think I've talked long enough and I'd like to leave some time for questions. Thank you very much.

*MODERATOR WILFORD:* Well, now we go back to gas again and Charles Neumeyer.

*MR. CHARLES NEUMEYER:* Well, Fred, you've talked quite a bit, you know, and you've covered both ends of the picture with respect to gas and electricity. I'd just like to comment that it seems to me it was not too long ago when you'd have a panel and have the panel members up here representing the oil industry, the gas industry, and the electric industry and the whole direction of the panel would be quite different. We would all be up here arguing as to how good our fuel was, and pushing our fuel to be used in preference to the others, and we would all be making a pitch to increase our sales or our load. However, here we are in a situation where we are all telling you why we don't have enough energy to serve you, so it has been quite a shift that has taken place, and it seems to me in a relatively short period of time.

Fred mentioned some of the things that I was going to talk

about, but I will just ramble on here a little bit. It seems to me that the emphasis at the general session this morning seemed to be on restricting any additional growth, and we heard the zero growth concept. Now it's very true that in some areas we do have this growth problem and that it has been a problem with respect to the electric industry. However, as you heard Chairman Swidler say, they have already imposed restrictions on additional utilization of gas in New York State and there are quite a few state jurisdictions throughout the country where any additional use of gas is now prohibited. So that as far as the gas industry is concerned here in New York State we're practically at a zero growth situation now. There are some exceptions to this prohibition on taking on additional loads, certain essential or necessary uses, but for all intents and purposes we're at that zero growth situation now. Our concern with additional sources of supply or new sources of supply is to a great extent to maintain our service to our existing customers. As was mentioned this morning, one of the big problems that we have is that our existing pipeline suppliers are curtailing their deliveries to us under the existing contracts that they have. This has upset some of our planning considerably because we didn't expect to be curtailed at all, or at least not to the extent that we have been with respect to the supplies that we had under contract.

Looking down the line in the short term, it looks to us that in order to bolster our supply or get additional supplies of gas, we have really two main new areas. The first one, of course, is going to be the Atlantic continental shelf and another source will be the Gulf Coast, through leasing of additional acreage out there. We think that there should be an acceleration of the leasing of this acreage and that the terms under which the acreage is presently leased should be changed. As you probably all know, the acreage in the Federal lands is put up for bids, and the bids are decided upon the basis of the magnitude of the cash bonus. Along with that bonus is a fixed royalty. This usually results in the fact that substantial monies are spent for the acquisition of the acreage and then there are not usually enough dollars left for rapid development of the acreage; that is, the drilling is usually postponed for a fairly long period of time. These leases are for a minimum term of five years and the leases can then be extended if there is any indication of activity taking place on them.

We believe that the continental shelf lands have to be explored more rapidly than before. Let me respond to Joe Kearney's question as to why we don't look at alternatives before we push for the development of offshore. Well, the on shore possibilities are being exploited now and there have been some restrictions on the exploitation by reason of the cost, as Joe mentioned. But when we consider the potential reserves still in the ground that are estimated by various groups, we come up with a number something like 1200 trillion cubic feet, or a little bit less than that. One of the things that we have to recognize is that over 300 trillion of that is in Alaska and not readily available to the continental states. That would leave something over 800 trillion as available to the continental states, and of that 800 trillion, a substantial amount is offshore. I believe it's something in the order of 200 trillion of that 800 trillion is offshore. Of the gas that's estimated on shore most is at fairly deep depths so that the technology involved in drilling and the costs are quite different from the kinds of activity that have been pursued on shore in the past. Offshore to us then is one of the most important sources of new gas.

Looking again at the relatively short term, the other new sources of gas that we are looking for are LNG, which as we mentioned this morning is liquefied natural gas from foreign sources, or the manufacture of substitute natural gas. These supplements are going to be necessary regardless of how much activity we have in drilling for natural gas under ground. All of the projections that have been made down to 1985, no matter what assumptions you make with respect to how much domestic natural gas can be got out of the ground, show that the requirements are more than the amount that can be obtained. Therefore, liquified natural gas from foreign sources and the manufacture of substitute natural gases are all going to be essential to meet the requirements.

Getting back to another question that Joe brought up as to what we need to do to get some of this gas that's in the ground flowing. You have heard that one of the problems has been the restriction on price to be paid to the producers and that this has had a dampening effect upon the exploration and drilling activity in the U.S. As a result of this dampening effect, the capital that's necessary for exploration has gone to other areas of the world. So what we do need is more realistic well head pricing of the natural gas, particularly the new supplies that are necessary. This is a fairly complex subject, and there are many differences of opinion as to how this should be accomplished. The Federal Power Commission has been criticized over and over again for the way it has lagged in coming up with new pricing to get additional gas flowing. However, the Commission has been in a bind to some extent because of all the activities that it has attempted. There have been questions raised with respect to the price justification. But this is something we do need. We need better price at the well head and we need some imaginative ideas as to what this pricing should be.

If we get better pricing at the well head, if we get the accelerated leasing of the continental shelf lands, if we are able to import sufficient quantities of liquefied natural gas from foreign sources, if we are able to get the capital and get the necessary permits to build substitute natural gas plants, we believe that we'll be able to take care of requirements for at least the short term, through 1985.

*MODERATOR WILFORD:* Thank you. I wish we had more time to get into this pricing situation because the only time I had it explained to me was by Texans who didn't look as if they needed more money. Now we'll have one last word from Robert Sampson of Cities Service Oil Company.

*MR. ROBERT SAMPSON:* Actually most of the story with regard to the demand in energy situation as it relates to the oil industry in projections through 1985 has been previously related. The one thing I'd like to point out to everybody, though, is that this is not an indication that we're running out of natural resources or mineral fuels to develop. It is simply that we are caught in a bind in supplying these fuels.

The energy gap in 1985 will be greater or lesser than current projections depending on decisions by government on all levels. Certainly our nation can narrow this gap if certain actions are taken. Due to the large lag time required in the development of natural resources, the time for action is right now. Our nation needs a national energy policy which would provide for both efficient development of fuel supplies and the wise and efficient use of energy supplies at a reasonable cost to the consumer and with adequate protection to the environment. The petroleum industry needs adequate returns on investments to attract the high risk capital needed to explore, develop, and do basic research for oil and gas and synthetic fuels. A decrease in real price of crude oil in the past ten years, rapidly increasing price regulations in exploration and development costs of natural gas, increased tax burdens including the Tax Reform Act of 1969, and the failure to allow the orderly exploration and development of new sedimentary basins have all contributed to this decline in exploratory drilling, the decline in oil reserves, gas reserves and the recent decline in the rate of return on investment for the domestic petroleum industry which dropped below 10 percent in 1970. This type of rate of returns will not attract the high risk capital needed for energy requirements in the 1970's and 1980's.

The price of natural gas should be unregulated and allowed to seek the price level of competitive fuels. The average price of gas in the United States is less than one third of the equivalent BTU value of oil. An upsurge of exploration for both oil and gas would result from this action. The Federal government should provide for the orderly development of Federal lands, including the outer continental shelf areas of this country. Additional oil and gas will no doubt be found in the producing provinces of the United States. However a major portion of the vast oil and gas reserves which remain to be discovered in the United States are in the relatively unexplored areas of Alaska, Alaska offshore, the west coast offshore, the Gulf coast offshore and the east coast offshore.

A large oil and/or gas find off the east coast of the United States would be the greatest insurance policy you could have here on the east coast for your future in environmental and economic needs. Energy is vital and the real challenge here is the development of ample field supplies and the protection and improvement of our environment.

I think this brief summary will give you a little better feeling for the oil supply picture.

*MODERATOR WILFORD:* Joe Kearney has asked for one word and since there are three industry men I'll let him have one more word.

*MR. KEARNEY:* Thank you very much. When we consider the cost of supplying our energy now, we must add in the environmental effect of each of the different alternatives. Only then can we evaluate the alternatives adequately.

I think that we ought to open the entire competitive market for the various sources of energy. If one company has requirements for a return on investment so large that it cannot produce oil at a competitive price, then it should get out of the business and let some other industry take up the slack until prices go up to the place where it can afford to produce again.

*MODERATOR WILFORD:* I understand now that there has to be one word from the audience.

*FLOOR:* I think there are two things to remember about the international oil situation. If we became too dependent on the overseas supply of oil, I think we would soon find that the price would jump, and that low-priced foreign crude oil would be a myth. The other thing is that the same crisis from which we are now suffering domestically as regards our oil supply will probably occur some time during the eighties on an international scale, and when this happens there is going to be severe competition for all available crude oil.

There is one other point I want to bring out. If you hold back the prices of crude, it will accomplish nothing but the dismantling of the domestic oil industry at a time when we are already suffering a serious curtailment of energy. Then our gap won't be ten million barrels a day, it will be twenty million barrels a day, and it would take a good ten years to reassemble a viable industry which could develop its own resources. Anyone who tells you what the import system is costing the American public is simply not considering all the other factors which should be added to the picture before a fair judgment can be made.

*MR. SAMPSON:* Let me say this much. From the standpoint of the industry, we truly believe that if you did away with the import quota system, we would see $5.00 or $6.00 crude oil in the country a lot quicker than you'll see it with the quota. Now there is no one today who is saying that we shouldn't import part of our energy supply or part of our mineral resources. It gives a certain degree of protection that our nation needs to augment our own ability to provide a quantity, especially during emergencies.

*FLOOR:* I find somewhat of a conflict in the recent decision of the FPC to permit the importation of Algerian and Libyan liquefied natural gas and yet to retain the quota system, since the liquefied natural gas is an obvious substitute for oil.

*MR. SAMPSON:* Well, we have got a shortage. The LNG is a substitute for natural gas here in the United States, not oil. As far as the industry is concerned, there is no problem as far as importations of LNG are concerned.

*FLOOR:* Isn't this a direct substitute for petroleum products when the gas is used for firing a utility plant and so forth?

*MR. SAMPSON:* What you're saying is that if you don't use

the natural gas, you're going to have to substitute oil for it. But what this natural gas will go for, primarily, will be residential and commercial uses. I agree with you that there is not much sense to a policy which holds well head prices and taxes at a quarter of a dollar and then allows oil or liquefied natural gas from Algeria to come in for a dollar and a quarter.

*MODERATOR WILFORD:* I hate to interrupt a solution of this problem, but we have about six or eight minutes until the resumption of the meeting downstairs, so unless someone has something very urgent that they want to say and they can say it in one minute we'll have to adjourn.

*FLOOR:* I understand from Mr. King of Texaco that only about 17 percent of the oil production domestically goes directly for the production of electricity, now that there are oil fired generators. I asked him what most of it goes for, and he said most of it goes for consumer uses in terms of gasoline. Now I think that the entire myth of the oil shortage in the United States which leads us to talk about importation of oil from God knows where is based on the automobile. If we could start to taper off our auto mania, we would find we would not have such a tremendous oil shortage in this country and that our oil production would not require importation. As far as the offshore goes, isn't this Federal land and therefore wouldn't it have to be developed with a large payment of royality to the Federal Government? I would just like to know who would get the revenue from the offshore oil.

*SAMPSON:* The United States Treasury.

*FLOOR:* You say there is a gap in choosing new methods such as nuclear fission, etc. It seems to me that any new generator could be built in a way that would satisfy the demands of all but unreasonable environmentalists. What the electrical industry is doing is riding roughshod over the legimate objections of the people in regards to safety, fish life, etc.

*MR. SULLIVAN:* We've been trying to build the Storm King Mountain project for 10 years and it is still in litigation. But I think this is really beyond this panel because there is another panel that is discussing the environmental balance situation. I wish we did have more time to discuss this particularly, but I think we still have to adjourn now.

*MODERATOR WILFORD:* Thank you very much.

# Workshop No. 2
# CRISIS IN ENERGY DEMAND

*Moderator*
Dean Lewis—WHN

*Federal Government*
Warren E. Morrison, Special Assistant to the Chief
Office of Economics, Federal Power Commission

*Environmental Protection*
Marc Messing, Executive Director
Environmental Planning Lobby

*City Government*
Martin Goldstein, Chief, Environmental Systems Planning
The City Planning Commission of New York

*Electrical Industry*
Edward J. Doyle, Jr., Vice President
Consolidated Edison Company of New York, Inc.

*Gas Industry*
James W. Dunlop, Manager, Rate Department
Brooklyn Union Gas Company

*Oil Industry*
Frank Young, Director, Economics Division
Continental Oil Company

## Workshop No. 2
## CRISIS IN ENERGY DEMAND

I. Overview
   A. Are our priorities/objectives compatible for action?
   B. National/international demand constrains city energy resources (competition for available supplies)
   C. Effects of differing demand breakdowns
   D. Institutional alternatives for energy management

II. Realistic Appraisal of Demand Growth Rates (through 1985)
   A. Population growth (inevitable modest demand expansion)
   B. Social side effect of energy constraint policy; standards of living among low income populations
   C. Additional energy for environmental protection (water treatment, waste management, removal of lead from gasoline, recycling)
   D. Air pollution and energy mixes

III. Conservation of Energy Resources
   A. Rationing and end use controls
      1. City authority/pre-emption
      2. Priorities
   B. Taxing certain energy uses
      1. Changes in rate structure
      2. Peak surcharges
      3. Tolls and City taxes
   C. Changes in life style
      1. Per capita use growth
   D. Improved energy efficiency
      1. New buildings
      2. Mass transit
      3. Appliance

*MODERATOR LEWIS:* Ladies and gentlemen, my name is Dean Lewis, I'm with WHN. And over the past several years that this has become such a burgeoning issue, I've been involved with the various problems that we face with our environment, directly as a citizen, and professionally in my capacity as a news commentator at WHN radio. I'll be serving as the traffic cop for this session of ours, trying to hold our discussion within some limits which have been defined by the title which has been given to this particular workshop, *Crisis In Energy Demand.*

I think perhaps the first step for us would be to define demand and let you know how our panelists here with me fit into this particular phase of the issue. So I'll introduce each of them, one at a time, and have them describe to us what they see as the problem of demand and tell you at the same time how they fit into this picture.

We'll begin with our friend from the Federal Government, Warren Morrison, who is the Special Assistant to the Chief Office of Economics, Federal Power Commission.

*MR. MORRISON:* The alternatives in any problem situation such as the so-called energy crisis are that we may have only several options open to us and both of these options I'm going to talk about are limited by a very important thing, the time factor.

The strategies of our energy problem in the short run are certainly going to be very much different from those required for the long term situation. In the short run, it's obvious that we can't do very much to change the energy system as it exists. And we're just going to have to solve our more immediate and pressing problems within this system. In setting our energy strategies then, we've really got to worry about two time zones. We've got to worry about the more distant future and this is where the government has the function. Certainly the government can change the energy system over a period of time by research and development and by emphasizing alternate conversion systems, prime movers, and so on, and also influence the environmental factors.

But over the short run, we're not going to be able to change technology. We're going to have to make our changes within the areas in which we are now working. Unfortunately, many of the problems that we're going to have to face are in this short term time zone.

Here and now, we have to deal with the problems of New

York City and environmental damage. We can't use the revolutionary technologies Mr. Swidler was talking about—the solar energy system, fusion and so on. So what are our real immediate problems in energy and those that we can really address ourselves to in the short run?

First of all, we've been faced with the continuing and aggravated imbalances between the demand for energy and the available supplies, particularly in primary resources such as natural gas, petroleum and clean fuels, such as low sulfur coal.

Secondly, the so-called exponential growth of the demand and the uses of these energy fuels is really having environmental effects that the nation just won't tolerate.

The social costs, Dr. Swidler pointed out, of remedying these past environmental effects and maintaining future environmental quality are just going to have to be borne. Within the limits of existing technologies then, what are our real options? First of all, we can increase the supplies of clean fuels through more imports. Dr. Swidler covered that. We can accelerate the rates of domestic development of what we know to be our assured reserves of clean fuel such as natural gas, and increase the generation of electric power. And there are limits to this, over the short run.

Finally, the other option is one that is more and more frequently brought up, and that is that we can conserve energy, and particularly electric power, by reducing the wasteful or inefficient or unnecessary uses of it. And we have other proposals which are brought up frequently such as absolutely reducing the demand for power, halting the development of offshore production, disallowing the Alaskan pipeline, deferring new power sources such as nuclear, and so on. I believe these to be counterproductive.

The government's role, as I see it, in the short run is to help regulate energy supply in the best, most efficient and most low cost way, and in the long run our role is really to help develop alternative energy sources.

*MODERATOR LEWIS:* All right. We won't let you get away from here without asking you how you intend to do that in view of the increasing demand.

Next we have the Executive Director of the Environmental Planning Lobby, Marc Messing, and I think the first question has to be what the Environmental Planning Lobby is, how did it come into being and how does it function?

*MR. MESSING:* Okay, the Environmental Planning Lobby is a statewide organization created to lobby environmental issues strictly on a statewide basis. Since Earth Day 1969, there have been some attempts on the national level for citizens' interest groups to begin lobbying Congress on environmental issues. Environmental Action, Friends of the Earth, and the Sierra Club have all been doing this, and the Environmental Planning Lobby is one of about six or seven attempts throughout the country to do this on a statewide level.

I think that's the briefest answer that covers the ground and if there are any further questions we can take them up later. Would you like to pose a demand question to me or can I take it from here?

*MODERATOR LEWIS:* Well, why don't we wait and see what demand questions our guests in the audience will have, Marc. Again, I want to emphasize that this will be a question and answer session and I hope it will be wide open and freewheeling so that you'll feel free to ask whatever questions concerning the demand locally that you would like to ask. Representing city government with the environmental systems planning of the City Planning Commission is Marty Goldstein.

*MR. GOLDSTEIN:* My area of responsibility is to make both short term and long term energy and fuel studies for the City of New York, and to act as a liaison with the Con Ed and various fuel people on how we can furnish electrical energy fuel, and also to make proposals to the City Government on how to conserve energy and fuel. We're also proposing to change various administrative laws in the city to facilitate conservation of various forms of energy. I think that, in a capsule, is what I do. If you have any further questions later, I'll be glad to answer them.

*MODERATOR LEWIS:* All right. It was not by specific design, but you'll notice that I have Marty well separated from Ed Doyle, Vice President of Con Ed on my left here.

*MR. DOYLE:* Thank you. Well, my end of the business at Con Ed is the commercial end. It was formerly known as marketing, and before that as sales. You might ask why I am still here and I think it is a good question, since at the end of 1970 we dissolved our sales department and discontinued all sales promotion and sales advertising, and introduced the Save-a-Watt activity. I'll just tick off a few of the things that I think are quite interesting in this. I'm still on the customer's side of this whole business.

As Mr. Swidler said in his very fine presentation, there's no way to restrict load growth without serious damage to our society. I wrote that quote down because I so agreed with it. The idea of the "frivolous appliance" also intrigues me because of the way some of these things are built up out of proportion. We will leave that for later. There are questions of air conditioning efficiency. We had an air conditioning seminar in our shop about ten days ago in which we had a large group of conservationists and environmentalists, consumer affairs people and the like. I'm interested in a number of the things that were mentioned this morning, some by Mr. Musicus, some by our President, some by others, questions about the building codes and what changes are needed, the sealed building idea, rate inversion, and various other ideas that I'm sure will come out in the questioning.

*MODERATOR LEWIS:* All right, and on the subject of rates, we now have, from the Brooklyn Union Gas Company, James Woody Dunlop.

*MR. DUNLOP:* I'm the forecaster for the Brooklyn Union Gas Company, forecasting the needs and demands for our customers for energy in the future. I have also been involved in some of the national forecasts so this is the expertise I bring. I can also perhaps answer some of your questions on inverted rate structures to the best of my ability.

*MODERATOR LEWIS:* All right, and representing the oil industry, Continental Oil Company, the Economics Division, Frank Young.

*MR. YOUNG:* I'm Director of the Economics Division of Continental Oil which is one of the major oil companies in other parts of the country. You're probably not too familiar with it here in the New York area. I've been involved in some of the National Petroleum Council studies on energy which are under the sponsorship of the Department of the Interior. Of course, I'm familiar with the demand, and I participate in demand studies on the oil industry, and as an economist I'm quite intrigued with some of the ideas that are being put forth here on how to come to grips with our problems. It seems to me that we've got about three things here that are of particular interest to me. We have something of a dilemma in society between increased energy use, which is put forth as necessary for economic growth, and energy needed to clean up the environment. This is, I think, an interesting aspect of our problem.

Also, as an economist, I'd like to put in a plug for reliance on a price system for making these decisions rather than having the city or federal government decide whether or not you can use an electric toothbrush or some other appliance. I think a price system is a much more effective and efficient regulator of this than government edict. I think another intriguing aspect of this problem on which we might want to focus attention, is the effect of rising prices on demand. As economists would say, as the price goes up, it should have some feedback effect on demand, again through the price regulating system of a free enterprise type of economy.

I think these are some aspects of our problems.

*MODERATOR LEWIS:* Well, you have led me, Frank, to what perhaps could be a kick-off question here that I could pose. And I would like to invite any of you to feel free to jump in here at anytime now. Just let me know when you have a question, an extension of the question I'm about to ask or one competely different of your own. And I would also like to encourage our panel members to participate in each others' discussions and answers. I don't mind arguments or disagreements in a situation like this and I think it might spark thoughts from some of the rest of us.

Do you feel that the price system is the way to regulate demand in view of our generally affluent society today, in which there are more and more people who are able to afford air conditioning or have air conditioning afforded for them? Frank, you've indicated that you do. Does that mean that you would support the inverted pricing system?

*MR. YOUNG:* Well, I think the price system certainly has a part in it, but I would concede that in some areas we need government regulation. A price system is probably not too effective in regulating demand, at least not in the crisis we foresee here, because most of the uses are tied to the original purchase cost of the automobile or the appliance. They're probably going to use the energy anyway, even though the price is increased. I think the price system may be more effective on the supply side, which is, of course, the subject of another panel.

I do think that it's still a useful mechanism. The public bodies should set the broad parameters, and then the price system can work within those. Otherwise, I think we would end up with a society which would be quite different from what we're used to, where we are told what we can and what we can't do. I don't think that's really compatible with our tradition of consumer freedom of choice.

*FLOOR:* What is the outlook for energy availability based on the NPC study? How does that match with Mr. Swidler's comments?

*MR. MORRISON:* Well, I think there is one basic assumption implicit in both of them. In the NPC study, it's implicit that you can find some method of stationary gas control so that you can meet the sulfur problem by using high sulfur coal. I think Mr. Swidler's assumption is that we will have this control and this represents the very basic premise on which their outlooks are based. I think it is still a matter of opinion, but a lot of research is going ahead on the problem and there have been some significant technological breakthroughs on the stationary gas problem. It is still too expensive, and has not got into the commercial area as yet, but a number of people who are quite knowledgeable in this field think that it will be commercially available in from three to five years.

*FLOOR:* How do you compare nuclear and fossil fuel plants for meeting energy needs?

*MR. YOUNG:* Well, there is, of course, a much shorter time span for putting in a fossil fuel plant than a nuclear plant. A nuclear plant will probably take about seven years now, whereas a fossil fuel plant could probably be built in about five years. Also, you can put a stat gas control for sulfur on existing plants. This is expensive to do; it's much cheaper to put it into the original plants, but it can be done. It will, of course, lead to higher costs for electricity.

*MR. MESSING:* Excuse me, I was eager to speak from the government side here. I'd like to frame this in a slightly different context. I think we all have to acknowledge the fact that we are in for a serious energy situation for the next ten, twenty, thirty years. There are going to be serious energy problems regardless of the specific form they take. As conservationists and environmentalists, we have received many requests over the past two years from people who write in and say, "What can we do to alleviate the energy shortage? Should we buy this or that? Should we unplug our electric toothbrush or should we turn off our air conditioners? What should we do?"

And to a significant extent, the answer that we have been giving, as responsible environmentalists, is a very cynical one.

We've told people that there's not a damn thing they can do as consumers. That they can throw out all of their trival electrical appliances and it's not going to do a thing to the energy projections. That they can choose a more efficient air conditioner rather than a less efficient one and it's going to have a very small effect. That, in fact, the factors which determine our electrical or our energy growth rates are embodied in premarket considerations. And that if we attack the premises of energy demand, we have to attack it at this premarket position.

In the premarketing elements, we've got price allowances which give certain benefits to the capitalization costs of production facilities. We've got a price system which is based primarily on the cost of delivering energy. That is why consumers decry the fact that large consumers get discounts whereas they, the small consumers, don't. They tend to overlook the fact that the pricing system is based solely on the cost of delivery, and that cost of delivery is less when you are delivering it in bulk.

And we've got a system which dismisses, or tradionally has dismissed, all of the social, the environmental and the economic external costs of energy production. Here in Con Ed the traditional definition of public interest has been that our role as a public utility company is to supply electricity whenever anybody asks for it. Not to tell them whether or not we will be able to supply it, or the cost that may be incurred in supplying it, but simply that if somebody wants to put in a building, we will supply them with the electricity.

*MODERATOR LEWIS:* Marc, let me interrupt here. I think it will be interesting to hear Ed Doyle's response to that particular point.

*MR. DOYLE:* Well, I think this goes back to one of the things that Swidler said. We've got a franchise obligation. We live up to that or somebody else is going to take over.

*MODERATOR LEWIS:* All right. That brings up the question then, is our government, are our elected employees in the seats of government, actually being responsive to the best interests of the community? Are these people meant simply to reflect our desires, or are they meant to be leaders in this field and perhaps by persuasion and from the public platform which they have, head us in a different direction than the one in which we're going now? That brought dead silence, didn't it?

*FLOOR:* Mr. Lewis, I think you've brought up a very valuable point about the leadership role that government has taken.

I think that the special problem with any dialogue like this is that government has laid a smoke screen over the issues. You take doctrines like the national power studies which assume that we're locked in, and say nothing about the possibilities of growth. I think the way that government and industry have defined the question is a way that says that nothing can be done. In other words, Swidler's report on the inverted price structure says that inverted price structures are going to have no effect in the short run, therefore they will have no effect in the long run. Or Mr. Young said that rates won't have effect on demand in any conceivable way.

*MODERATOR LEWIS:* Well now, just a moment. Are you proposing then that the total cost of production should also include the amount of damage being done to the environment?

*FLOOR:* That was point number one. Point two is the current structural rate. As far as I can tell from these studies on the rate structure of Con Ed, there's very little relationship to the other costs, the non-environmental costs.

*MR. MORRISON:* Well, one of the things about internalizing external costs is that the utilities would have to start to charge you, the consumer, for these internal costs on your bill so that the states and the Federal Power Commission could take this into consideration in their rate cases and allow higher rates.

Secondly, as far as the ability of a regulatory commission, whether it's a state commission or a federal commission, to really make changes depends upon changing its legal structure to begin with. The Federal Power Commission operates under the National Gas Act and the Power Acts. And we're restricted by what these Acts allow us to do.

*FLOOR:* What is the FPC doing to solve the energy problem?

*MR. MORRISON:* Well, if you would read the annual report of the Federal Power Commission to Congress which we make every year you would see that we have asked for additional authority to allocate gas and power if necessary. This is one thing. We're not allowed to do this under the present legislation.

I think the point of internalizing costs is that the consumer has got to face the fact that these costs are going to be paid by him. And I haven't really heard anybody up to this point saying that they're willing to pay these costs.

*MR. DOYLE:* I'd like to suggest that they're already paying in part. In our case, to bring the sulfur content of the fuel we burn down from one percent to point three percent is costing our consumers about $100 million a year, almost $10 million a month. Also, our rate base, on which the FPC decides what our rate of return should be against a standard or allowed rate of return, includes vast capital sums for changeover to oil burners and precipitators and the like.

*FLOOR:* What impact does price really have on energy demand?

*MR. YOUNG:* Our elasticity studies show that we've got a long way to go before price will influence consumption. Elasticity is an economist's word for saying what kind of a price increase you need on the supply side to bring forth the required supply, or what kind of a price shift you need on the demand side to change the demand. Now, quite a few Ph.D. dissertations have been written on one side or the other of this question.

On the inelastic side, they're saying that you can increase the price of energy to the American consumer by a very wide margin before you get him to change his patterns of consumption. On the supply side the Federal Power Commission has increased the price of gas at the well head by a considerable margin over the past year, and we haven't had any appreciable response on the supply side.

So this is really in a sense the crux of the problem, this thing called elasticity. If you're going to use the price as a lever, at what point do you have to increase it to get response, either on the supply or the demand side?

*FLOOR:* Why are you proposing inverting the price, causing the customers to pay more for electricity? Why can't the City change the Building Code to reduce energy demands? Is the City planning to stop new development?

*MR. GOLDSTEIN:* I think you don't understand what price inversion means. Price inversion means that the more of the unit you use, the higher the price goes, so that essentially a person who uses a minimal amount of a particular goods pays a base price. But as he uses more of the commodity the unit price goes up; this is a price inversion. Well, the price could be worked out so that a small consumer of electrical energy won't have any price increase in rate. However, a person who uses vast amounts of electrical energy in commercial buildings or industrial facilities, will be required to pay more per unit of electrical energy.

At the present time, you have a scale rate that says the more of a commodity you use, the farther down the cost per unit goes, and this is what we're opposed to.

Your second question was in regard to the building code. New York City is presently making a study of the possibility of revising the building code to include additional insulation on buildings to facilitate minimizing thermal loss through the walls and through the ceilings. We're also investigating the type of fenestration, the type of windows which are used in buildings. In fact only last week we met with various manufacturers of insulation and window equipment.

We're also proposing conservation techniques for using less energy in buildings. We've been meeting with the illuminating engineering society on changing lighting standards. Now what is unique, by the way, in this room is that the lighting standards are very low. They're perfectly satisfactory, but I would say they're something like 20 foot candles or something; it's a really nominal amount. You could actually manage if you turned off the electricity and opened the curtains. Behind these curtains there are windows, and just to have these lights off above here, you could probably save about 1,000 watts or so in here.

And what is the city doing about new development? In regards to Manhattan Landing and, for that matter, any new large scale development in the city, we're requiring the developer to make comprehensive plans on energy demand and possibly also energy supply. We have a liaison established with the person on the other side of the table here with Con Ed to alter their operations and possibly get into development of total energy or selective energy facilities for large scale development; that is, it's conceivable that Con Ed could build, on particular sites within the city, new energy plants, which by the way we hope would have a thermal efficiency of somewhere around 50 or 60 percent, instead on some of their older electric generating plants now where thermal efficiency is somewhere around 20 percent.

Now what that means is that the more efficient you are in your energy use, the less fuel you use, and also the less pollution you will discharge.

*FLOOR:* In talking about demand and the question of architecture, there is a question I want to ask. To begin with, we must realize that the architect who designed this building, the interior decorator who put the curtains in front of these windows, did so to achieve certain functional ends and certain esthetic ends. This is the problem that style brings to this area. I am thinking of the dilemma of the consumer, the basic dilemma which our whole society faces when confronted by the promotional genius of America which is designed to increase consumption. Is there any activity going on to reduce consumption?

*MR. DOYLE:* I'd like to make a comment here. We talked about the "frivolous appliances." But these are not the big consumers. Even lighting does not cause the big loads. When you consider the whole energy picture, the big loads are caused by space heating, space cooling, water heating, automobiles, mass transit, sewage disposal, water pumping. We have tremendous growth in those areas going on in New York City. Just in the area of broad services to the public, we are proposing the extension of the transit system, and so on.

We need to conserve energy in the field of heating and cooling of space, and that's where matters of insulation, matters of reducing the glass exposure, and so on, come into the picture. There's a lot to be done in that line, and we've been doing some work on it ourselves.

*MR. MESSING:* I think I can answer that question a little bit further and I'd like to.

*CHAIRMAN LEWIS:* You have about 30 seconds to do it.

*MR. MESSING:* The state of Connecticut has imposed a mandatory building code and considerable work has been done on that by the AAAS and by Oak Ridge National Laboratory.

*MODERATOR LEWIS:* I'm sorry, the time that we have had allotted here has gone far more rapidly than it should have for all of us I'm sure. I'm going to have to decline your question because we're due over on the other side now. So we are going to have to adjourn. Thank you.

## Workshop No. 3
## ELEMENTS IN THE ENERGY-ENVIRONMENT BALANCE

*Moderator*
Frank Field, NBC-TV

*Federal Government*
Kenneth Johnson, Director, Air & Water Programs
Environmental Protection Agency

*Environmental Protection*
Fred Hart, Commissioner
New York City Department of Air Resources

*Electrical Industry*
Robert O. Lehrman, Vice President, Public Affairs
Consolidated Edison Company of New York, Inc.

*Gas Industry*
Eugene Luntey, Executive Vice President
Brooklyn Union Gas Company

*Oil Industry*
William K. Tell, Jr., Associate General Counsel
Texaco Inc.

# Workshop No. 3
# ELEMENTS IN THE ENERGY-ENVIRONMENT BALANCE

I. Overview
   A. New energy reserves must be developed efficiently without unreasonable environmental delays
   B. Ten year planning lead times for developing new resources
      1. New exploration preceding widespread shortages
      2. Timely environmental impact studies/early public discussion
II. Impact of Inadequate Energy Supplies
   A. Economy/GNP—unemployment, investment inhibiting
   B. Insecurity/Reduced standard of living
   C. Slow-down/Abandonment of public sector programs
   D. National security
   E. Environmental protection: air pollution, water and waste management
III. Energy-Environment Balancing
   A. Air pollution
   B. Offshore drilling
   C. Increased tanker traffic and reliance on fuel imports: Oil and LNG
   Nuclear
      1. Radiation and waste disposal
      2. Interim nuclear licensing
   E. Plant Siting (operating, refining)
      1. Air quality
      2. Thermal pollution
      3. NYS Power Plan Siting Act

*MODERATOR FIELD:* My name is Frank Field, I'm your friendly weatherman. You must thank me for this bright sunshine, it's a little hazy, there's a little air pollution out there but nevertheless it's a nice day. I was here yesterday and got caught in the rain so I have no complaints.

I'd like to get very briefly from the members of our panel some statements as to their stand and their feelings, I'd like this to be an informal, open give and take even amongst yourselves, the more disagreements the better on the subject. So let me begin now with our first panelist, Mr. Kenneth Johnson, the Director of Air and Water Programs of the Environmental Protection Agency.

*KENNETH JOHNSON:* Thank you and thanks for the weather. I'm going to talk briefly more or less in a philosophical sense about control in our environment and how it relates to the production and use of energy. What we consider the basic Federal or environmental protection agency programs with regard to air pollution control right now are in great portion based upon air quality standards and plans designed to achieve and maintain those standards.

The standards are of two basic types. Primary standards which are designed to protect public health, and secondary standards which are designed to protect the public welfare from any known or anticipated adverse effects of air pollution.

Now the first point that I want to make here is that these standards are based on already demonstrated effects upon people and their property from pollutants in the atmosphere, so in essence we're reacting to an already existing problem.

Secondly, the pollutants that we're talking about, sulfur dioxide, particulate matter, oxides of nitrogen, carbon monoxide, hydrocarbons, photochemical oxidants, all of these by and large, especially in a Metropolitan area like this one, are related to the combustion process both for the production, some times for the use of energy.

So what we really have here is a case where pollution control in the atmosphere is really a reaction to already existing effects of the use of energy, and the burden of proof, even now, is on the public to demonstrate the effects of the control action that's taken.

The result is that we have a built-in conflict with other priorities. For example, in the City of New York, we need more natural gas in order to achieve the primary or health related standards for amount of particulate matter allowable in the air by 1975.

However, there are other people throughout the country who are using natural gas for other reasons, such as its cheapness, and naturally they are not willing to give up their priorities. It's a constant fight on the part of the public of New York City who wish to use natural gas because they want to protect their environment, to try to get priorities. It is more difficult because they are coming in after the fact, and struggling against already established uses.

The same thing applies in the production of power. Even if we disregard the frivolous uses of power, it's an acknowledged fact that power generation is going to go up in the future. But people who are there to protect the environment from the effects of combustion products associated with the generation of power say, "Let's not produce the power here in the heart of our city, but somewhere else." Then we're told that this is where all the people are who have to use the power, and so this is where we have to produce the power in order to take care of them in the near future. Again we're facing a case where people who are working for controls and protection of the environment are coming in after the fact.

In the case of the automobile industry where the Federal government is asking the automobile industry to control the emissions of carbon monoxide and hydrocarbons from their exhausts by 1975 and oxides of nitrogen by 1976, the automobile industry says, "Well, we can't do that. You're going to cripple us, so give us a break and give us more time." Again in almost every case it's the environmentalists who have the weakest side because the great balance of existing priorities are against them.

I don't think that this can continue. It's a case of necessarily having to look ahead and deciding a long way in advance that our highest priorities are the air we breathe and the water we drink. We must plan for the production and use of energy in advance to guarantee that we don't have to have the environmentalists in the future take a back seat. And you may say that it's all right to talk about the future, but that we're moving toward that right now, and I think we are.

I think that the trend of citizen interest in governmental action throughout history has been to ever move toward this position of planning in advance, taking the burden of proof to demonstrate the effects of pollution away from the public and putting it on the polluter. And I'll cite two cases in which we're starting to take those steps now and I think they foretell the future.

In the first case the National Environmental Policy Act of 1969 with regard to any significant federal action, whether issuing a permit or actual construction itself has ruled that an environmental impact statement has to be written which addresses itself not only to the specific action which is being undertaken but to the uses for the project which is to be built, so that eventually we are going to have to look ahead and plan, and that's a good thing for the future.

The second thing is the Cleaner Air Act Amendment in 1970 in which one of the requirements is that the States in their implementation plan have to discuss and have the authority for land use control designed to achieve and maintain the air quality standard.

But here we're talking about a basic feature of planning: the ability to control land use for protection of the environment. I'm convinced myself that this is the way of the future; it won't simply be a case of the public reacting after the fact, demonstrating the effect upon themselves or their property and saying, "Help us out of our suffering."

*MODERATOR FIELD:* Thank you, Ken. Our next panelist is Fred Hart who is the Commissioner for the New York City Department of Air Resources.

*FRED HART:* I think I must react in some way to the remarks that were made at lunch time today. I think, as Ken points out, that it's an energy-environment balance and I would add to that a third factor which is resource. We're really faced with three problems: energy need, resource capability and environmental need and it would appear from this afternoon that the choice according to Mr. Swidler's speech would be energy first, resources second and environmental needs third. I think this is an unfortunate turn and in an unfortunate direction.

I think I need to say that because the objectives that we have in New York State and New York City to meet the very standards which the Federal government has established as far as air quality is concerned require a complete reversal, which is that the environment be considered first.

The approaches which we have taken and which are being taken throughout the country are in considerable conflict to those taken in New York State.

There are at the present time nine or ten states in the country that have a power siting organization which has responsibility for locating power plants.

In all but two of them (one of those two now being New York State), the power, the ultimate decision maker in the process is an individual who represents the public, who does not represent the industry. I think it is important to get this out on the table at this time as we begin to discuss this particular problem.

It is appropriate to consider the issue of land use in the State and also in the City. When you consider the energy problem, the resource problem, the environment problem in New York City you must as well consider the use of fuel in heating and providing hot water in the City.

We now believe that it's possible in New York City to actually have environmental benefits by centralizing plants. If we were to substitute one central plant which provided steam, for example, we would have a favorable environmental tradeoff with the 5,000 or so oil burners that would be replaced.

It's possible at the same time that we develop a plant to produce this steam to also provide electric generating power. This is a proposal that individuals within the City Administration have proposed for the so-called Manhattan Landing project.

Clearly after we get involved and after we search out and develop our priorities, we would all agree that one of the areas in which we must move ahead is research, and I think that the research must begin in the resource area.

It's very apparent that one of the major constraints on energy is not the environment but really the resources which are available. It's also very clear to those of us who have had some exposure to the Appalachia area that the coal resources which are available there might find an appropriate application in urban areas such as New York City.

The efforts which Consolidated Edison Company are making in the desulfurization area at their Arthur Kill plant are an appropriate beginning. We think that there are a number of other areas for the properly modified use of coal in New York City. We think that research efforts should be devoted in this resource area.

*MODERATOR FIELD:* Thank you, Fred. Our next panelist is Mr. William Tell, Jr., General Counsel at Texaco. Mr. Tell, would you like to throw in just a few words?

*WILLIAM K. TELL:* Thank you, Frank, I have just three or four comments that I would like to make at this point in the program. I think the speakers this morning, and certainly our luncheon speaker, a man who brings to us a tremendous background and experience in the energy field, made it quite clear that the country and the city are facing critical energy shortages and that the lead time has run out and that the time for talk and theory is also running out and we're at the decision point and we must act unless we want to find some very disruptive and painful side effects entering our society.

Therefore, I would hope that a conference such as this today could mark the entry into phase two of the energy environment struggle, that we try to find a dialogue and try to get away from the polarization that has marked the problem in the past three or four years with environmentalists talking among themselves, industry all to often talking to itself, neither side perhaps fully appreciating and understanding the concerns that the other was feeling and trying to express. Because time is gone and now we must do something.

Now, lead time is one part of the problem; another part of the problem is trying to find procedures and institutions that will best help us find solutions.

I would have to take exception to the procedure which Mr. Sive suggested this morning, which as I understand it was continued litigation, continued advocacy, a continuation of the adversary process.

I think that's too slow, I think that's too emotional and I don't think we have the luxury of continuing on a case by case basis. I don't think the judiciary of the country has the time or the expertise to provide us with solutions to these very complex technical, scientific, political and economic problems. I think the solution is going to have to be found through legislation, and the quick adoption and implementation of a coordinated national energy policy.

There seems to be a recognition that this is necessary; there are some encouraging signs in Washington but the progress has been all too slow. Too often the step that is taken is only the proposal of another new study. Studies are fine, but when the red lights are flashing on every side we just simply cannot responsibly say that the step which needs to be taken now is another study .

I think we need to find concrete, constructive solutions to the problem. Now, one thing that I would like to offer as a suggestion, specifically to the future energy requirements of New York

City, is a close look at proposals from the Federal government that the Atlantic coast, outer continental shelf be opened for off-shore leasing.

This is an area that offers good prospects for the discovery of substantial new reserves of oil and gas. It would be the next logical frontier in which the petroleum industry could concentrate its efforts in trying to keep up with the energy of the country.

It is an area unlike the West coast where operations are conducted only three or four miles offshore. Because of the gentle slope of the Atlantic shelf, the areas of interest would be thirty or forty miles offshore, out of sight, out of hearing and with an adequate area for containment if there were the unfortunate incident of a spill.

The petroleum industry would bring to such a project a past experience of over 16,000 marine wells. This is not a novel area of operation or activity for us; in that experience of 16,000 wells we have had only one incident where there was serious pollution on the shore line, which was at Santa Barbara. The studies that have been conducted at Santa Barbara since the time of the spill in 1969 show that the area has enjoyed a full ecological recovery.

I think this is a subject that is going to have to be looked at very closely by New York State. It's a subject that has generated considerable controversy on Long Island, but I think when we're trying to recognize that now we must find a balance, we will find that the solution does not lie at either polar extreme.

We will recognize that offshore drilling statistically is a safer way of getting energy than by increased tanker movements, that the geology of the Atlantic coast offers good prospects for finding new quantities of natural gas which this area needs so badly, and which could make a great contribution to improving the quality of the city's air. When a balanced judgment is made we will see that this is a positive step that should be looked on favorably by New York City and its residents.

*MODERATOR FIELD:* Thank you, Mr. Tell. Our next panelist is Mr. Eugene Luntey, Executive Vice President of the Brooklyn Union Gas Company.

*EUGENE LUNTEY:* I think you've probably had enough speeches for today and I think that all of us on the panel are far more interested really in your remarks at this time. I would like very briefly to go over the position that Brooklyn Union has taken in this environmental field for a good many years. Ken and I were reminiscing up here and neither of us could quite believe that it was five years ago in 1967 that the Federal hearings on air pollution started in New York City. At that time I testified that natural gas could do more to alleviate the air pollution problems in New York City than any other solution that anyone could come up with.

That same statement goes today, but the progress that has been made in supplying gas to New York City is very, very minimal. Actually there's hardly any more natural gas being used in New York City today than there was in 1967 and this is because of a number of factors.

Brooklyn Union is not completely free of blame in this since for about a 10-year period, from the mid 1950's to the mid 1960's, Brooklyn Union fought in the Federal Power Commission hearings almost every increase in the price of gas.

Now there was a reason for this. At that time there was a surplus of gas being produced, and the Federal Power Commission was in the middle of hearings to try to cost base the regulation of natural gas on the same basis that a utility has to justify their rates on a basis of cost. The oil companies are not free of blame either, because in that litigation that went on for so many years they did not fully understand the utility concept and did not fully include all of the future costs which we're beginning to have to pay today.

Now that's all water over the bridge but the problem with natural gas today is continuing delay on the Federal Power Commission level. This delay is because some environmental groups do not quite understand what the problem really is with natural gas, and the New York State legislature just passed a bill which supposedly prohibits drilling off the state of New York.

Now that bill is ambiguous and it's ill advised, and yet there were very few senators or representatives when this came out on the floor who had the courage to ask what was really being done.

As Bill said, natural gas produced off the East coast can do a great deal for the every day problems of the people of New York.

It seems to me that we've got two problems to discuss today, one is a short term problem and that is what can we do right now? And that's the place we need a great amount of cooperation from everybody, because our aims are the same. We have no desire to pollute the atmosphere or deface the land in which we live, but we do have an obligation to furnish energy to the people who are our customers.

The near term problem I believe can be in an area of complete cooperation with every environmental group here or in the country.

The long term problem, which is one in which we have to design energy systems that will allow this country to live in harmony with the environment for the next 20, 30, 50 years, includes time for the type of confrontations, litigation and long term procedures that Mr. Sive was talking about this morning. The near term does not; the problem is here now, there is a 15% curtailment on natural gas firm contracts that are being supplied to New York City at the present time.

The increase in particulates in this city which has been experienced this last year, I believe to be at least partly influenced by that curtailment which started this last winter.

So I'd like anyone here to approach this on that basis, that we do have a near term problem and that we need cooperation and that we need to know how you think we can approach this problem that we have this year and next year, and then in the long run what are we going to end up with as an energy system for this country?

*MODERATOR FIELD:* Thank you Mr. Luntey. Our last panelist is Mr. Robert Lehrman who is Vice-President of Public Affairs of Consolidated Edison Company of New York.

*ROBERT O. LEHRMAN:* I think there's been enough mention of Consolidated Edison this morning and through the lunch hour to warrent my not saying anything at all, but since we are talking of energy in the environment I'll just mention a couple of problems which have been referred to this morning and try to bring them down to some pariculars.

I think it's particularly important to recognize that electricity is but a small part of the total energy problem; electricity represents about 25% of the total energy that's utilized in this country. Of course electricity is awfully visible, and Con Edison, patriotic as it is, makes it even more visible with those gorgeous red, white and blue stacks that surround us here in New York City.

I'd like to mention that we're also a gas supplier, a large one, the fourth or fifth largest gas company in the country.

First I would like to mention this concept that has been proposed (many of you may have seen it in the newspapers) of mini-power plants here in New York City. Some propose them as a solution to our energy and to our environmental problems; some people think that they may be a useful step towards a solution of the problem.

The concept that was described most recently in the press related to a gas turbine generator on top of the roof of a large development such as the one down in lower Manhattan. This generator would produce electricity and the waste heat would be taken from that gas turbine and be utilized to supply heating and perhaps cooling in the development. And someone suggested that this would be a useful thing to do in all other large developments which are under construction here in the city.

We think that this is a very worthwhile thing to study, and we are studying it. We also think that as with any of the other solutions that have been proposed, many of which been discussed here whether they be solar energy or fusion or the rest, there is no easy answer.

Let us take this proposal as a kind of a case study. Now we know that gas is in increasingly short supply, which means that these so-called gas turbines would probably have to be run by fuel oil and that means of course, combustion to produce electricity.

It means also many units, separately controlled, located in different places in the city as contrasted with large central units which, of course, are more easily controlled, and better controlled from the point of view of the environment.

It means oil trucks going through the city to supply them, or digging up the city streets to put the necessary lines in to supply them.

So from an environmental viewpoint, obviously there are questions that must be examined before there is any wholesale acceptance of this particular concept as a solution to our power problem in the city.

In the point of view of reliability, these mini-power plants obviously would need some form of backup, that is some form of reserve of capacity to supply the large development if for some reason that particular mini-power generator were to go out of service.

And lastly, of course, and this must be considered when we're trying to deal fairly with the consumer, we must face the question of cost. Is it in the economic interest of the consumer as well as in the environmental and power supply interest of the consumer to have numbers of small plants with all that entails in terms of cost, as contrasted with a larger centrally located power plant, most probably as we have agreed with the city, outside of New York City.

So the solutions to this question of electricity and the environment are not easy ones, as we've heard this morning.

Even the question of central station steam is not an easy one; consider the question of the famous or infamous East 60th Street Sutton Place steam plant, which is in some substantial controversy at this time. The city itself supports the concept of central station steam as an environmentally desirable way of supplying energy. And yet there are many difficult questions which come up when we are trying to apply that concept in any individual case, as we have found in the case of that plant which is still in the midst of controversy on environmental grounds.

That concludes my brief remarks; I'd be pleased to respond to questions. We are in the midst of a troublesome time in trying to resolve the conflicting interests of the environment, resources, power supply and of cost to the consumer. Thank you.

*MODERATOR FIELD:* Thank you, Mr. Lehrman. I'm sure if you've listened to the speakers you agreed wholeheartedly with each one of them in little aspects and disagreed violently in other directions and I could stand up here and ask questions all afternoon and maybe get some good stories out of it, but I think we ought to open up the floor now and see whether speaker versus speaker on the podium here or any of you folks out there have questions. We'll begin right down front.

*FLOOR:* Mr. Lehrman, in New York City I believe the average person contributes about five lbs. of garbage per day, and there is a problem in picking this up and disposing of it, burning it in set incinerators around your city or dumping it out at sea. What is the potential for compacting this garbage and mixing it with some other fuel and installing any necessary cleaning devices in your stacks as a supplement to our other fuels that we use?

*MR. LEHRMAN:* The question relates to the use of garbage and waste as a fuel which could be combusted to produce electricity.

First, Con Edison has problems with waste too. Our Arthur Kill generating plant out on Staten Island is troubled frequently by the fact that there is a spill from the Fresh Kill garbage dump, the waste from which comes into the water and then gets pulled into the power plant. Of course, the amount of power we are able to get out of the power plant is reduced at such a time.

The question of large scale use of garbage as a fuel in power plants is another of those questions the answer to which is not going to be with us immediately. It is being done to some extent in Europe, but it's being done in rather small installations. It is being done in a couple of places around this country, but again in small installations. The problem is to make waste into a useful fuel and at the same time not to have air pollution problems associated with the burning of this mass of miscellaneous material.

It probably would be extremely difficult if not impossible in terms of cost and also maintaining power supply to install such capacities to burn garbage in our existing power plants. It may well be possible on newer units and, as a matter of fact, we have talked in our 20-year power plan about the possibility of a power generating station on the site of the Sing Sing Prison at Ossining on the Hudson River, and we've mentioned in connection with that a possibility, perhaps as a pilot or a demonstration, of utilizing some waste as part of the boiler fuel for that facility.

Again it's one of those proposals that's not the answer but something that should be explored in the overall interest of the community.

*MR. JOHNSON:* Environmental Protection Agency within recent months has been concerned with the very problem that you have raised with regard to the Hackensack meadowlands across the Hudson in New Jersey. What had been proposed there and may be still proposed, is building the world's largest incinerator, a 6,000 ton per day unit to produce heat for 150 megawatts of power. The Environmental Protection Agency considered this proposal and looked at other alternatives and issued a report in March which reflected essentially the position of what used to be the National Air Pollution Control Administration which was that if there were other non-combustion ways to get rid of refuse, they would be the preferred methods. One of the reasons for this decision is that not only might we be involved with the common air pollution factors like oxides of nitrogen or particulates from the combustion, but that more exotic combustion products might cause still greater problems.

For example, the Hackensack meadowland incinerator we're talking about might perhaps produce in excess of 5,000 tons per year of hydrochloric acid from the burning of chlorinated plastic resin. It might also emit various heavy metals and materials which accumulate in the body and are toxic at high concentrations. If, however, for one reason or another it's essential to incinerate and there's no other way out, then we do believe that the practical use of heat for the generation of electricity is worthwhile, because then at least you don't use as much coal or oil or other fuels to produce the electricity.

I believe in the long run that we probably would be better off with a system like pyrolysis, which would essentially boil off the gases and then use and control those and reclaim a lot of materials and eventually work toward a full resource recovery and recycling program and not simply try to get rid of the refuse by burning it up.

*MR. HART:* There's only one point to add, I think, and I would agree that it would be quite difficult to apply this type of technology to existing Consolidated Edison plants. I think that it would be feasible for the relative medium term, let's say the five to seven year time in new installations. Another consideration that should be brought in is the cost of collection. If it were designed including a vacuum system or something like that which would dramatically reduce the manpower to pick up garbage so that collection costs were much lower, the savings might be profitably spent in good air pollution control devices on these particular units.

*MODERATOR FIELD:* Thank you, gentlemen. Two of our panelists have come up with a suggestion that one can drill offshore and get gas and possibly alleviate this problem. Now I'm not sure whether this is for the short range outlook, or long term outlook. I'm sure that there are those of you who disagree violently with that idea; why don't we have an exchange on that?

We really are all agreed we're all for research; we're all looking for ways to improve what's happening to our atmosphere; we're all in agreement that we don't want anything adverse to happen to that atmosphere or to our water or to our environment.

The question now is, if we are to believe what's been said this morning, that we've come to that point where there is no return, we're going one way or the other. Many people give me the impression that they do believe we've got to start making a move. Mr. Tell has mentioned we can't take these things through court, case by case, something has to be done. What do we do? Is there a scientific way out? Must we give up some of our rights to the environment in terms of energy?

Or should we curtail energy and suffer that way and enjoy the environment? What do you think?

Let me make it a little stronger, put a little more pressure just so we can get something going here. Have we reached the point in our scientific life that we can glibly say, "Well, we'll contain that accident"?

It seems that Thor Heyerdahl, going over the Atlantic saw oil all over the place. You simply can't scoop it all off the top of the ocean. Can we assume that we are working on this problem?

*MR. TELL:* Well, I'm happy to talk on that point. If you'll recall I said that out of 16,000 wells there was only one incident where there had been serious pollution on the shoreline. There have been some incidents in the Gulf of Mexico, but they really are only a handful, I mean ten at the most over a period of perhaps twenty years where there has been a blowout. As a result of the new techniques that the industry has learned to utilize before and since Santa Barbara the recent incidents in the Gulf were handled differently. There was a Chevron fire and an Amoco fire. If you'll recall, the industry on those occasions, rather than sending Red Adair and the fire fighters in to dynamite the hole and put the fire out that way, which results in some quantity of oil being spread over the area, engaged on a very costly program of drilling a series of relief wells around the platform that was on fire. They let the fire continue to burn even though that caused very serious damage to the platform and caused the companies to accept quite an economic penalty, but as long as it was burning, the oil and the gas were being consumed by burning. There was no pollution on the waters.

In those instances as a result of drilling those relief wells, which snuffed out the blowout well, they were able to contain the situation without any of the oil getting to the shoreline.

Now, Frank, you asked whether we could say that there would never be an incident. I don't think that anyone would believe me if I said there never will be an incident, because we're operating equipment with human beings who are fallible. I think what we're talking about today is finding an accommodation and a balance, so that we can find some real solutions to some tough problems.

Now there are few things in life that involve men and machines that do not carry with them some risk, at least theoretical, of a malfunction, which could cause undesirable side effects of one form or another. These are things which are a fact of life that we have learned to accept, live with, and attempt to minimize in many areas of our human activity.

Now because of the emotion that has been associated with certain aspects of our environmental movement, I suggest to you that perhaps in the past we have attempted to impose standards which realistically were not attainable, and did not allow for viable solutions to the problem. What we say to you is that we think we can develop oil and gas reserves off the Atlantic Coast, that we can do it in a responsible way with a risk to the environment that will be far less than if you continued to look toward increased tankers and imported energy as filling your needs, and which will be far less than the risk that you will accept in another area of your life in terms of air quality.

We think if you look at the record coldly, objectively, and statistically, you would conclude that when you balance the side effects of inadequate energy resources against what theoretical risk may exist in terms of Atlantic Coast offshore development that the judgment would have to fall on the side of developing the outer Continental Shelf reserve.

*FLOOR:* Mr. Tell is approaching the crux of the question, which is how you achieve a balance if you have an oil spill, whether it goes to shore or remains out at sea, against the need for resources? How can the methodology that's used there be applied to the problems such as those Consolidated Edison might have in the City of New York. I wonder if Mr. Johnson might comment on that in terms of perhaps a cost benefit analysis?

*MR. JOHNSON:* Well, I think that when we look at solutions to the problems in New York City for getting gas in order to achieve standards of air quality by 1975, certainly we should look at other sources of natural gas, certainly we should look at the possibility for exploring off the coast and drilling. At the same time we should look at other means, too.

For example, I believe that it's technologically possible that you can take crude oil itself and produce naphtha and from the naphtha produce methane for use. You can bring in by ship sources of liquefied natural gas. All of these and perhaps more should be examined to determine which is the most satisfactory means for taking care of the immediate problem, and which has the least adverse impact upon the environment in every case. We're looking at this ourselves right now, to choose the one that has the least cost to the people.

There is talk about the use of natural gas on an accelerated or expedited basis in certain target areas of the country where the air pollution problems are the most pressing as in New York City, rather than using it all across the entire country. We're looking at the possibilities of using a dual fuel system, an interruptable system which allows for the use of less natural gas in the times of greatest consumption, at a cost that people can pay. We shouldn't use one single solution and say, let's aim at this right now and get it; we should examine all of them.

*MODERATOR FIELD:* Well, this takes us into research and the long term planning. I would like to take you back again to the immediacy. When you step outside there's air that you must breathe and there's water that you must bathe in and drink. One obvious question I would like to ask Mr. Tell as a reporter is: assuming we grant you that privilege that you so eagerly seek, how long would it take? How long would it last?

*MR. TELL:* That gets into the question of lead time which we think again is the kernel of the problem to a country that has never known energy shortages. It's perhaps tempting to take a "hope springs eternal" attitude that there will be some radical breakthrough and that we won't have to do anything painful now. But unfortunately if we wait until we're in a period of shortage before we apply the medicine, and that's what we've done as far as natural gas is concerned, the lead times can be as long as seven to ten years and that's going to mean a period of very painful side effects.

Now to be specific in terms of the development of the Atlantic Coast outer continental shelf, there are several things that must occur before the industry could even commence its activities. One of them would be environmental impact hearings where there would be an opportunity for all interested parties to participate and examine seriously what the implications would be.

You may have read in the papers that over the past three or four years the industry has been conducting geophysical activities off the Atlantic Coast. This is not drilling; these are techniques which permit a better understanding of the subsurface geology so that when appropriate government clearances are available the industry will be able to identify those portions of the Atlantic Outer Continental Shelf where the subsurface structures would appear to offer the best potential.

Now fortunately we already have that phase fairly well along, but in terms of doing the basic exploration drilling, after hopefully identifying a field or a structure that contains reserves, then it becomes necessary to enter into a phase that is known as developmental drilling. That's when additional wells are drilled around the reservoir to permit the most effective production of the subsurface reserves. Platforms would then have to be constructed to support the producing effort once the field was developed, pipelines would have to be constructed to bring it ashore and these activities would require before the full rates of production could be achieved from offshore, somewhere between five to eight, perhaps even ten years after the Interior Department has granted the leases.

*MODERATOR FIELD:* Now how much could we expect to gain in terms of time on that, assuming that this immediately happens tomorrow? Assuming that you went full steam ahead, and you got all the reserves, do you have any idea of how much material there is that is desirable under the ocean and how long it would sustain us?

*MR. TELL:* Now in terms of the quantities of reserve, no one could make precise forecasts until wells were drilled. We are able through geophysical techniques to have some understanding of the subsurface geology but until the bit goes into the ground and goes into the reservoir you do not know whether there is a commercial well there or a dry hole.

Now statistically in the United States you are looking at a success ratio of perhaps one in 37 exploratory wells finding hydrocarbons in commercial quantities. So even though you have some understanding from your geophysical techniques, you still have to drill before you can say how much is there and whether it is oil or gas.

The only drilling on the Atlantic Coast has been off Nova Scotia in the Sable Island area within the last year or so and that did have several indications of natural gas, which is encouraging. That would be the greatest thing that could happen to New York City if substantial supplemental reserves of gas could be found.

Now how long it would last once we found it? The normal depletion period again would depend on the rate at which you produce the wells. You would want to produce it at a rate that would reflect conservation principles and not prematurely dissipate the pressure in the reservoir and leave most of the oil and gas in place. That was the trouble at the turn of the century before we had a better understanding of subsurface mechanism. But I would say that the life of a commercial field operated under prudent conservation principles might span a period of twenty to thirty years.

*MODERATOR FIELD:* Thank you. Well, you know what the odds are on that.

*FLOOR:* Aren't we actually leading ourselves up to a short range solution by the importation of liquefied natural gas? You realize that we don't like to commit ourselves to importation and reliability on the Mideast, but I do know that Algeria is extremely anxious to sell and develop some of their resources so they can get the large cash to develop their countries. We're talking about a twenty year period that will give us the chance to develop enough new sources of energy.

*MODERATOR FIELD:* This would be carrying what form? Liquefied?

*MR. LUNTEY:* The problem with LNG as far as a large volume solution is concerned, is that it requires so much capital on both ends and in between. Really it's because it requires a large, very expensive liquefaction plant in the Algerian end of of this business, very special ships which are tremendously costly, and then it requires storage facilities on this side because you can't use it as rapidly as it's delivered by the ships. Brooklyn Union is starting to import LNG this next winter. We are to get two a half billion by barge from Everett and then the following year the Staten Island Terminal should be ready. But the cost of that gas is about $1.60 compared with about 50c for standard pipeline gas and not only that but it has to be limited in quantities because of the tremendous capital investment required in between.

So, as far as the solution is concerned, this is part of the solution, this will do for taking care of the peak points, the unusual period in the winter time, but I don't see it as a large volume solution. There are other important offshore developments—the North Sea development, the development off Norway, the Holland-Belgium development where those countries sat for so many years on large reserves of natural gas without ever knowing about it, but development off the East Coast is important because of its location close to the Metropolitan areas of the

East coast and we really cannot afford not to explore it. It's far safer, in my opinion, than the tanker traffic which is really the only feasible alternative to it.

The other part of the offshore development that I'd like to speak about for just a moment is that we believe that the air pollution effect of New York City has a far greater effect and does far more harm to the marine atmosphere than would any possible spillage from the worst platform accident that you could imagine.

This air pollution is unique in the sense that on the West Coast the air pollution generally drifts inward because of the prevailing wind that flows from the west. But the air pollution off the East coast generally drifts out to the marine atmosphere and day after day has a tremendous effect.

Now there has been very little research done on this, but it's one of the factors that needs to be considered in this entire balance of the environment.

*MR. JOHNSON:* Mr. Luntey, when you say that the cost of liquefied natural gas will be $1.60 I assume that's on a firm basis?

*MR. LUNTEY:* I should say that this $1.60 is not the price being quoted, on a firm year round basis. This is gas that has to be stored for about half the year because the use of gas is such that it's needed only during the six winter months. The LNG imports will come in at somewhere around $1.20 on a firm year-round basis, but then you have to add to that a cost of storage, in our case for the winter use and vaporization.

*MR. TELL:* If you had liquefied natural gas that was used on an interruptable basis, could you give an estimate of the cost there?

*MR. LUNTEY:* I don't think we'd use LNG on an interruptable basis. It's just too expensive. You would not use it as a base gas and then supply oil, for example, for the rest of the period, it's just too expensive.

*FLOOR:* What is the status of the air pollution implementation plan? Has it been approved by the Federal Government? Did it include the use of natural gas in automobiles?

*MR. HART:* In the plan that was submitted to the Federal Government was an amended plan that had been prepared by New York State. I do not know whether or not it has been formally submitted or whether or not it has been signed by the federal government.

This puts that strategy near the bottom of potential alternatives. There was appended to the New York State implementation plan a series of recommendations from New York City which did not include that particular strategy, mainly because of the very strong objections of the New York City Fire Department.

*MODERATOR FIELD:* All right. Has anyone any comments?

*MR. LUNTEY:* Let me just comment on what we've run up against in some hearings in Nassau county in trying to drill 30 to 300 miles off the coast of New York.

One of the theories that was propounded, without any or with very little geological evidence, was that such drilling off the coast of Long Island would cause an earthquake because of a fracture which was said to run the length of Long Island Sound.

Now this argument is competely without foundation in my opinion. There is no evidence that drilling has caused earthquakes, certainly not in the Gulf of Mexico where there has been tremendous drilling nor in any other area of the world that I know of. This is the type of argument that would make drilling quite difficult and might delay that type of development for a number of years.

*MODERATOR FIELD:* Thank you. I'm afraid that this is the time at which we must come to a halt. Thank you, members of the panel and audience. The meeting is adjourned.

## *Summary Session*

**CRISIS IN ENERGY SOURCES AND PRODUCTION**
John Noble Wilford
*The New York Times*

**CRISIS IN ENERGY DEMAND**
Dean Lewis—WHN

**ELEMENTS IN THE ENERGY-ENVIRONMENT BALANCE**
Frank Field, NBC-TV

**CLOSING REMARKS**
Marian Heiskell, Co-Chairman
Council on the Environment of New York City

*NEIL ARMSTRONG:* I have not introduced Mr. Allen Smith of Brooklyn Union Gas who is one of our Directors and actually was the individual who conceived of this idea of utilizing this format to bring everybody together. Allen, will you take a bow.

There were a lot of questions that were not answered this morning, and I want to assure you on behalf of the entire committee that we want a specific answer for all questions wherever possible.

If any of you have additional questions we urge you to write either directly to the individual or the panelist or to the Board of Trade, and we can assure you of an effort to answer it and furthermore, we welcome your suggestions and recommendations as to a continuing dialogue, and how it can be most effectively developed and utilized. We thought this might be Chapter One, but maybe this is only the prologue to what we hope will be a very interesting book.

At this time it is my pleasure to introduce Mrs. Marian Heiskell who is known to many of you here for her complete dedication to making New York a better place to live. Her latest endeavor has been on terms of the environment, and her work has been most effective and deeply appreciated. So at this time I would like to ask Mrs. Marian Heiskell to give us the summary for the afternoon.

*MRS. MARIAN HEISKELL:* Thank you, Neil. I must say I think the Council on the Environment is indeed grateful to the Board of Trade for allowing us all to get together here, and it's been very successful.

I'm going to do this a little bit backward, I'm going to start with Workshop Three which was headed by Dr. Frank Field, because he has to get back to tell us about the weather. He covered elements in the energy and the environment balance.

As you well know, Dr. Field is the science editor and meteorologist for National Broadcasting System; he reports on the six o'clock and the eleven o'clock news; he also moderates research projects. His Sunday half-hour television program has won numerous awards including one from the Medical Society of New York County.

Dr. Field holds numerous science degrees including meterology, geology and optomology degrees.

He is on the staff of the Albert Einstein College of Medicine and serves as an advisor to many organizations in the New York area. Dr. Field, would you try to summarize your workshop?

*DR. FRANK FIELD:* Thank you. The reason I'm coming on first is that I've got to get back to the studio, which is three blocks away, before it rains.

Many years ago I was involved in air pollution research with Dr. Leonard Greenberg. At that time people were mostly concerned about their own welfare, including the smoke stack directly across the street or the smell coming from up the block. There was no interest in one's fellow man. I'm delighted that we've come a long way since then.

Now our panel dealt with the energy and environment balance, I think the panelists all agreed, as did our group, that there was an imbalance and that we had to do something about it. Practically everyone agrees we've come to the point where there is no return, that we've got to take some steps.

Frankly I was disappointed in a way because there was not much more argument. I expected a little more fire. Mr. William Tell of the Texaco Company and Mr. Luntey of the Brooklyn Union Gas company offered a suggestion as to how we might help our environment by offshore drilling and I expected to see some chairs come out of the audience, but I'm delighted to say that there is a dialogue, everybody sat back. They may have been stunned, I'm not sure, but we did have a very nice dialogue. No agreement, you wouldn't expect that. The key question was how long it would take to get it into the works, if it were a solution. How long would that solution, if it worked out, help us through this crisis? There are no answers to these questions until the actual work. It was a commendable suggestion.

Mr. Hart, who is Commissioner of New York City Department of Air Resources, addressed himself to general problems and the improvements that are underway today and also pointed out that what we need today is more research in the area of resources and that we should be improving the methods of getting the energy and in turn that would help us in terms of our environment.

Mr. Johnson brought out the important fact that up to now it has been the consumer who has had the burden of proof placed upon him and that that has changed. There are now real steps that are being taken on the Federal and State levels that mean an improvement; he was rather optimistic about it.

Mr. Lehrman came out with his idea that possibly instead of having fragmented sources of power, the best approach was to get one centralized source of power and energy which would do away with a great deal of the pollution.

Now personally I think all this is commendable, but we really skirted the main issue, which is, how much are we willing to give up on either side of the fence? Are we willing to give up on the energy so we have a better atmosphere, better water? Or, are we willing to accept lower standards in order to have more and cheaper energy.

Now, I don't think we can address ourselves to that; we can only offer some suggestions. We did that, and I'm not sure we answered very much of what was on everyone's mind. How far do we have to go in either direction? I think that's about it. Thank you.

*MRS. HEISKELL:* Thank you, Dr. Field. Next, we have Dean Lewis who moderated the panel, "Crisis in Energy Demand." He is a news commentator for WHN radio with news and commentary. His program runs four times daily, so you can definitely not miss him. He goes on at 7 and 8 A.M., at noon, and also at 7:05 P.M. He has offered broadcast and commentary for 20 years this month; we congratulate him for having lasted that long. Mr. Lewis is a citizen of New York and active in many civic affairs. Mr. Lewis, will you tell us about your discussion?

*MR. DEAN LEWIS:* We had a rather interesting time, I thought, in our Panel. Again we came to no conclusions except perhaps that the crisis in demand is pretty much an individual matter for individual consumers as well as for individual industries and companies and at this point it would seem we were left with the feeling that none of these individual entities are yet ready to say, "I'll be first to begin curbing the demand."

One of the problems I noted immediately was that the government agencies, who were given the responsibility of dealing with demand seemingly do not have the power to themselves begin putting limits on the demands. Con Ed is obligated to meet whatever demand there is, therefore, it cannot limit the demand.

Someplace along the line, therefore, it appears to me, we must set up a mechanism that can begin putting effective curbs on the frivolous ends of our demands for energy, if indeed we have finally decided that the production of energy in all of its various myriad forms is harmful to our environment to such a point that we must curb it.

On that point I don't think there was much disagreement,

but that is about the only point. We decided right away that there is a serious energy crisis, that took us very little time at all.

It was suggested that perhaps the government should take the lead in helping us to define the question of exactly what the frivolous ends of this demand are, what can be cut out and what must be kept.

It was suggested, and I think generally agreed to by everyone at the meeting, that one way or another customers do pay the total cost of the energy crisis as well as the pollution that is caused in the production and in the use of the energy.

There was no agreement on the proposal that perhaps an inverted price structure would help to limit the demand. As a matter of a fact it was pointed out by one of our panel members that indeed the increase of price and control on the producer and the increase of prices for consumers in some areas has done absolutely nothing to curb the demand for energy.

Contrary to what Frank mentioned a few moments ago in relating that one of his panel members had suggested that we centralize our power sources, we discussed more fracturing or fragmenting of the present generating resources in the City to provide energy—electrical energy specifically in this case—from localized plants spotted throughout the community, perhaps in the areas of large new developments.

There was also a discussion of the steps being taken by groups of architects, and the desire on the part of the City to have architects take these steps, to design new buildings in such a manner that there will be a reduced loss of heat in the winter and cooling in the summer, and better natural lighting which would reduce the need for artificial lighting such as we have in this room now.

Again there was no agreement, there was no solution offered as to how we, at this point, intend to curb or reduce the demands made upon our energy resources.

I must agree with Frank, and I think this is vital, that as he pointed out at least we have begun a dialogue. We have got together to begin discussing these problems—all of us who are concerned as private citizens, as environmentalists, as power and energy producers. This is a giant step forward from what could have been considered even as short a time as two years ago. I, for one, am very pleased to see us take this step. Alan, I congratulate you and the rest of the people who have made this possible and I thank you for allowing me to be a part of it.

*MRS. HEISKELL:* Last but not least is John Wilford, who is science writer for *The New York Times*. He started his career down in Tennessee writing a column for a paper called *Parisian*, in Paris, Tennessee.

Then he went to *The Wall Street Journal*, and after a stint in the Army did some free lancing in Mexico and then was a contributing Editor to *Time* magazine. Finally, in 1965, he joined *The New York Times*.

He's been in the science department ever since, covering the aerospace program. Last July he won an award for his series on the nation's energy crisis. John, will you tell us about your panel?

*JOHN WILFORD:* Our discussion went very smoothly until a few controversies arose at the end. We limited our discussion quite severely just so that we could try to get some substance into an hour's meeting. We limited the discussion of energy sources and production to what is feasible for the New York area in the years between now and 1985—these are the crisis years that are already upon us.

It seems, from the discussion by Mr. Wakefield of the Department of the Interior, that we have two choices. Because of the limitations on our discussion Mr. Wakefield eliminated certain possibilities for the more distant future, such as the breeder reactor, which will probably not be commercial until the 1985 to 1990 time period. Geothermal and tidal energy sources are promising but of limited possibility both geographically and in terms of availability. Solar energy and thermonuclear fusion are very long term aims and will not play a part in our problems between now and 1985. Also the oil shale reserves that we heard much about, the tar sands out west, were something for the period beyond 1985.

So he narrowed it down to two choices. One is to import more oil if we want to keep on supplying the demand here in the New York area and the east coast. Projections by the government are that if this course is taken we probably would be importing 57% of all our oil by 1985. This could create many problems for our already precarious balance of payments and it also raises some international issues, so it is not a very enticing prospect by his assessment.

The other choice is to develop more domestic oil and gas. And here we get into the question that I gather the other panels got into—the business of bringing the drilling derricks east, away from the Pacific and the Gulf, and introducing them out off Long Island somewhere.

Mr. Wakefield stated that there are four objections usually raised to offshore, outer Continental Shelf drilling. One is the general one of esthetics. Who wants to be sitting off Montauk point or Cape Cod and look at a derrick? Well, his answer there is that both of the derricks would be beyond 30 miles at least, and therefore over the horizon.

The second objection is the fear of disaster. And here there is no assurance that there wouldn't be another Santa Barbara off the coast of New York. However, he points out there have been only ten incidents in the last 35 years, and only about three of them were considered of a real disaster nature, and that now there are greater restrictions being imposed.

The third objection is the threat of greater pollution in the ocean. He quoted Coast Guard statistics as saying that only 2% of the oil pollution in the ocean water comes from the offshore drilling and the 29% comes from tanker spills and cleaning out tanker bilges. He points out that this is one argument against relying more heavily on importation of oil as opposed to offshore drilling.

The fourth objection is that this might interfere with fishing off the coast and he cites the fact that fishing in the Gulf of Mexico was actually improved over recent years.

In regard to the time when this would be possible, he estimates that from the time when authorization to drill and explore was granted it might take two years to the discovery of what was really there, and five to seven years before there was full development. In other words, if we started now it would be five to seven years before we could perhaps be pumping oil and gas out of the Atlantic.

On the other hand we had Joe Kearney of the Natural Resources Defense Council who discussed the fact that there are perhaps more alternatives to the problem. He discussed the necessity for improved efficiencies in electricity production and suggested that perhaps we should rely more heavily on nuclear power once we see the results of the safety questions now being debated in Washington.

He also made the suggestion that for the really short term, since New York has a higher peak in the summer in the consumption of electricity whereas New England has a higher peak consumption in the winter, there should be a greater tie-in between the New England and the New York grids to share the peak loads.

I don't think Mr. Kearney, as the environmentalist on the panel, was necessarily opposed to the offshore drilling, it's just that perhaps we were rushing this question a little too fast at this point.

One question he raised which was raised even more decisively later, was the question of the import quota. He suggested that we reduce or eliminate the import quotas to help supply some of our needs at this time.

We also had representatives from Con Edison, from Brooklyn Union Gas, and from Cities Service Oil Company on our panel. The Brooklyn Union Gas Company representative, Mr. Neumeyer, was a strong proponent of more drilling offshore in the outer Continental Shelf. He also emphasized the possibility of a greater importation of liquefied natural gas and a greater reliance in the future, as the technology develops, on synthetic natural gas. He said this is going to be necessary even if there is new drilling offshore.

Fred Sullivan of Con Edison pointed out that we burned our last coal in one of their generating plants in February 1972. Hopefully the next time we will be using coal will be in exotic mixes from the coal degasification processes. But between now and 1977 the real growth in Con Edison's capacity would have to come from nuclear plants and from the plants that are already authorized or under construction.

Robert Sampson of the Cities Service argued that one way we could increase the source of natural gas and oil would be to improve the rate of return for the exploration and development of more fields in the country. He stated that at the present time the rate of return does not attract the investment needed in the future to develop new sources. This is true in research work as well as exploration. He suggested that the Federal government should provide for the orderly development of Federal land including the outer Continental Shelf and Alaska in order to get the oil and gas that we'll be needing between now and 1985.

So, as one panelist summed it up, it is not at this time a resource question, but it is a supply question for the very near future. Very often it's an economic question too.

*MRS. HEISKELL:* Thank you. I've been allowed to say the last word and that's not very often allowed in my household.

As we finish today I sense the unsettled feeling that comes

with a difficult beginning. Nonetheless, I'm firmly convinced that the cities' needs and energy needs in a struggling environment will be balanced. If not through reasoned action now, by less than reasoned public necessity when the shortage hits.

For today the essential point is to begin to understand. I'm more worried now than I was this morning, but I'm also much better informed. In that vein an observation from an old Pacific coast Indian pertains to the close of our conference on energy and city environment.

In watching the weather and the lighthouse the old fisherman said, "Lighthouse him no good, excuse me. Him no good for fog; lighthouse him whistle, him blow, him ring bell, him flash light, him raise all kinds of hell, but the fog come in just the same."

We dare not wait for the fog to come in on the energy situation. Now is the time for getting on with it. The city needs an energy policy born of judgment rather than a byproduct of crisis.

Even though we are pressed, we still must inspect the issues, elicit public opinion and provide for the environment for the city's ability to do work. We may not be as exhaustive or as extravagant with our policy making as in the past, but we can be just as clear about what we can consider and what we cannot.

Some promises have been struck already in the cause of New York City energy. The Indian Point licensing here is an example. For the sake of city energy we do need timely action, but not to the exclusion of reasoned and adequate environmental protection in the process.

I see the plight of institutions in active noncommunication often. I have the insightful position of being on both the Board of Consolidated Edison and also the Co-chairman of the Council on the Environment for New York, so I've seen how company and regulator can talk busily at each other with neither side really listening.

They've each got different objectives, different approaches and different allegiances.

From here forward we need patience and active listening, we need to be able to speak each other's language and so define real direction for action, and we all need each other's good faith to proceed in the spirit of mutual effort to solve the New York City energy problem before it's too late.

Perhaps there isn't a right answer in the traditional sense.

Perhaps the question is more that of balance of public interest; after all the benefits are everyone's as are the costs of city energy.

Should not this public interest also have paramount weight in the balancing too? Popular awareness is the possible solution, which might mean the difference between support and apathy.

In perspective the energy issue may be the tempering of the environmental crusade, a kind of loss of naiveté in the realities of decision making. In a sense the new approach is for the environmentalists to join the business communities in getting down to distasteful specifics, and the mutual undertaking of the new order of public responsibility. This environmental realism will test all our good intentions and show what we really mean through what we actually do. As has often been said today, questions should not go unasked, for the sponsors of today's conference stand committed to continue in liaison between all participants, for further exchange, deeper inquiry and a policy for action. Your written questions will be answered by mail.

If time did not allow an answer today let us know how you want to go forward; we'll help and help the city, too.

Now let's get on with it and stop whistling in the fog. Thank you and thank you all for coming here.

# SELECTED READINGS FOR "ENERGY IN THE CITY ENVIRONMENT"


**A. Federal Government**

*Goals and Objectives of Federal Agencies in Fuels and Energy,* Senate Interior Committee Print, 1971.

*Selected Readings on the Fuels and Energy Crisis,* House Interior Committee Print, 1972.

*Energy, the Ultimate Resource,* House Science Committee Print, 1971.

**B. City Government**

Hallman & Fabricant, *Towards a Rational Power Policy,* Report from Environmental Protection Administration to the Mayor's Interdepartmental Committee on Public Utilities, April, 1971.

Mayor John V. Lindsay, Untitled Speech to Association of Home Appliance Manufacturers, New York City, May 3, 1972.

**C. Energy Industries**

*Consolidated Edison's 20-Year Advance Plan,* Consolidated Edison Company of New York, Inc., 1971.

E. H. Luntey, *Statement on Gas Shortages,* Brooklyn Union Gas Company, March 9, 1972.

"Offshore Drilling Issue", *Petroleum Today,* American Petroleum Institute, 1972-1.

**D. Environmentalists**

*Oilspill,* Sierra Club Battlebooks, 1971.

*Energy,* Sierra Club Battlebooks, 1972.

D. E. Abrahamson, *Environmental Cost of Electric Power,* Scientists Institute for Public Information Workbook, 1970.

**E. Professional Community**

"Energy to the Year 2000", *Technology Review,* MIT, 1972.

*"Electricity and the Environment",* Bar Association of New York City, Summer, 1972.

# STATE OF NEW YORK

## 9800-B

# IN SENATE

## March 7, 1972

Introduced by Messrs. McGOWAN, BLOOM, B. C. SMITH—read twice and ordered printed, and when printed to be committed to the Committee on Public Utilities—committee discharged, bill amended, ordered reprinted as amended and recommitted to said committee—committee discharged, bill amended, ordered reprinted as amended and recommitted to said committee.

# AN ACT

**To amend the public service law, the public authorities law, the condemnation law and the public health law, in relation to the siting and operation of major steam electric generating facilities**

*The People of the State of New York, represented in Senate and Assembly, do enact as follows:*

Section 1. The legislature hereby finds and declares that there is at present and may continue to be a growing need for electric power and for the construction of new major steam electric generating facilities. At the same time it is recognized that such facilities cannot be built without in some way affecting the physical environment where such facilities are located, and in some cases the adverse effects may be serious. The legislature further finds that it is essential to the public interest that meeting power demands and protecting the environment be regarded as equally important and that neither be subordinated to the other in any evaluation of the proposed construction of major steam electric generating facilities. Without limiting the generality of the foregoing, the legislature finds and declares that under certain circumstances power demands may be regarded as controlling even though the adverse environmental impact may be substantial, but that under other circumstances, given the nature of the resources involved and the public interest in preserving and enhancing the quality of life, the protection of the environment may be regarded as controlling even though this might result in restrictions on the availability of public utility services.

The legislature further finds that the present practices, proceedings and laws relating to the location of major steam electric generating facilities are inadequate to protect the environmental values and to take into account the total cost of society of such facilities and result in delays in new construction and increases in cost which are eventually passed on to the people of the state in the form of higher utility rates. Furthermore, the legislature finds that existing provisions of law do not provide adequate opportunity for individuals, groups interested in conservation and the protection of the environment, municipalities and other public bodies to participate in a timely and meaningful fashion in the decision whether or not to locate a specific major steam electric generating facility at a specific site. The legislature therefore hereby declares that it shall be the purpose of this Act to provide for the expeditious resolution of all matters concerning the location of major steam electric generating facilities presently under the jurisdiction of multiple state and local agencies, including all matters of state and local law, in a single proceeding in which the policies heretofore described shall apply and to which access will be open to citizens, groups, municipalities and other public agencies to enable them to participate in these decisions.

The legislature further finds that there is a need for the state to control determinations regarding the proposed siting of major steam electric generating facilities within the state and to cooperate with other states, regions and countries in order to serve the public interest in creating and preserving a proper environment and in having an adequate supply of electric power, all within the context of the policy objectives heretofore set forth towards which objectives the provisions of this legislation are directed.

§ 2. The public service law is hereby amended by adding thereto a new article, to be article eight, to read as follows:

## ARTICLE VIII

### SITING OF MAJOR STEAM ELECTRIC GENERATING FACILITIES

Section 140. Definitions.
141. Certificate of environmental compatibility and public need.
142. Application for a certificate.

§ 140. Definitions. Where used in this article, the following terms, unless the context otherwise requires, shall have the following meanings:

1. "Municipality" means a county, city, town or village in the state.
2. "Major steam electric generating facility" means a steam electric generating facility with a generating capacity of fifty thousand kilowatts or more.
3. "Person" means any individual, corporation, public benefit corporation, political subdivision, governmental agency, municipality, partnership, co-operative association, trust or estate.
4. "Board" means the New York state board on electric generation siting and the environment, which shall be in the department of public service and consist of five persons, one of whom shall be the chairman of the public service commission, who shall serve as chairman of the board, one of whom shall be the commissioner of environmental conservation, one of whom shall be the commissioner of health, one of whom shall be the commissioner of commerce, and one of whom shall be an ad hoc member appointed by the governor, who shall be a resident of the judicial district in which the facility as primarily proposed is to be located. The term of the ad hoc member shall continue until a final determination has been made in the particular proceeding for which he was appointed. Upon receipt of an application under this article, the chairman shall promptly notify the governor. Three of the five persons on the board shall constitute a quorum for the transaction of any business of the board, and the decision of three members of the board shall constitute action of the board. In addition to the requirements of the public officers law, no person shall be eligible to be an appointeee of the governor to the board who holds another state or local office. No member of the board may retain or hold any official relation to, or any securities of, an electric utility corporation operating in the state, nor shall the appointee have been a director, officer, employee thereof. The appointee of the governor shall receive the sum of two hundred dollars for each day in which he is actually engaged in the performance of his duties herein plus actual and necessary expenses incurred by him

in the performance of such duties. The chairman shall provide such personnel, hearing examiners, subordinates and employees and such legal, technological, scientific, engineering and other services and such meeting rooms, hearing rooms and other facilities as may be required in proceedings under this article. The department of environmental conservation shall provide associate hearing examiners. Each member of the board other than the appointee of the governor may designate an alternate to serve instead of the member with respect to any particular proceeding pursuant to this article. Such designation shall be in writing and filed with the commission.

5. "Certificate" means a certificate of environmental compatibility and public need issued by the board pursuant to this article.

§ 141. Certificate of environmental compatibility and public need.

1. No persons shall, after July first, ninteen hundred seventy-two, commence the preparation of a site for, or begin the construction of, a major steam electric generating facility in the state without having first obtained a certificate of environmental compatibility and public need issued with respect to such facility by the board. Any such facility with respect to which a certificate is issued shall not thereafter be built, maintained or operated except in conformity with such certificate and any terms, limitations or conditions contained therein, provided that nothing herein shall exempt such facility from compliance with state law and regulations thereunder subsequently adopted or with municipal laws and regulations thereunder not inconsistent with the provisions of such certificte. A certificate for a major steam electric generating facility may be issued only pursuant to this article.

2. A certificate may be transferred, subject to the approval of the board, to a person who agrees to comply with the terms, limitations and conditions contained therein.

3. A certificate issued hereunder may be amended as herein provided.

4. This article shall not apply:

a. To a major steam electric generating facility if, on or before July first, nineteen hundred seventy-two, an application has been made for a license, permit, consent or approval from any federal, state or local commission, agency, board or regulatory body, in which application the location of the major steam electric generating facility has been designated by the applicant; or if the facility is under construction at such time;

b. To a major steam electric generating facility the construction of which has been approved by a municipality or public benefit corporation which has sold bonds or bond anticipation notes on or before July first, nineteen hundred seventy-two the proceeds of or part of the proceeds of which are to be used in payment therefor;

c. To a major steam electric generating facility over which any agency or department of the federal government has exclusive jurisdiction, or has jurisdiction concurrent with that of the state and has exercised such jurisdiction, to the exclusion of regulation of the facility by the state;

d. To normal repairs, replacements, modifications, and improvements of a major steam electric generating facility, whenever built, which do not constitute a violation of any certificate issued under this article and which do not result in an increase in capacity of the facility of more than fifty thousand kilowatts; or

e. To a major steam electric generating facility (i) constructed on lands dedicated to industrial uses, (ii) the output of which shall be used solely for industrial purposes, on the premises, and (iii) the generating of capacity of which does not exceed two hundred thousand kilowatts.

5. Any person intending to construct a major steam electric generating facility excluded from this article pursuant to sub-division four may elect to waive such exclusion by delivering notice of such waiver to the chairman of the board. This article shall thereafter apply to each major steam electric generating facility identified in such notice from the date of its receipt by the chairman of the board.

§ 142. Application for a certificate. 1. An applicant for a certificate shall file with the chairman of the board an application, in such form as the commission may prescribe, containing the following information and materials:

(a) a description of the site and a description of the facility to be built thereon, including available site information, including maps and description, present and proposed development, source and volume of water required for plant operation and cooling, and as appropriate, geological, aesthetic, ecological, tsunami, seismic, biological, water supply, population and load center data;

(b) studies, identifying the author and date thereof, which have been made of the expected environmental impact and safety of the project, both during its construction and its operation, including a description of (i) the gaseous, liquid and solid wastes to be produced by the facility, including their source, anticipated volumes, composition and temperature, and such other attributes as the commission may specify, and the probable level of noise during construction and operation of the facility; and (ii) the treatment processes to reduce wastes to be released to the environment, the manner of disposal for wastes retained and measures for noise abatement; (iii) the concentration of wastes to be released to the environment under any operating conditions of the facility, including such meteorological, hydrological and other information needed to support such estimates; (iv) architectural and engineering plans indicating compatibility of the facility with the environment; and (v) how the construction and operation of the facility, including transportation and disposal of wastes, would comply with environmental, health and safety standards, requirements, regulations and rules under state and municipal laws;

(c) estimated cost information, including plant costs by account, all expenses by categories, including fuel costs, location plant service life and capacity factor, and total generating cost per kilowatt-hour, both at plant and including related transmission, and comparative cost of alternatives considered;

(d) a statement explaining the need for the facility including (i) reasons that the facility is necessary or desirable for the public welfare and is not incompatible with health and safety; (ii) the load demands which the facility is designed to meet; (iii) how the facility will contribute to system reliability and safety; (iv) how the facility conforms to a long-range plan for the development of an integrated statewide power system;

(e) a description of any reasonable alternate location or locations for, and alternate practical sources of power to, the proposed facility; a description of the comparative advantages and disadvantages of each such location and source; and a statement of the reasons why the primary proposed location and source is best suited to promote the public health and welfare, including the recreational, and other concurrent uses which the site may serve; and

(f) such other information as the applicant may consider relevant or as may be required by the commission or the board. Copies of the application, including the required information, shall be filed with the commission and shall be available for public inspection.

2. Each application shall be accompanied by proof of service, in such manner as the commission shall prescribe, of:

(a) a copy of such application on

i. each municipality in which any portion of such facility is to be located, as primarily proposed or in the alternative locations listed. Such copy to a municipality shall be addressed to the chief executive officer thereof and shall specify the date on or about which the application is to be filed;

ii. each ex officio member of the board and on the chairman of the board for transmission to the appointed member as soon after his appointment as practicable;

iii. the attorney general;

iv. the director of the office of planning services;

v. each member of the state legislature in whose district any portion of the facility is to be located, as primarily proposed or in the alternative locations listed;

vi. in the event such facility or any portion thereof, as primarily proposed or in the alternative locations listed, is located within its jurisdiction, the Hudson river valley commission;

vii. in the event such facility or any portion thereof, as primarily proposed or in other alternative locations listed, is located within its jurisdiction, the St. Lawrence-eastern Ontario commission.

(b) a notice of such application on

(i) persons residing in municipalities entitled to receive a copy of the application under subparagraph (i) of paragraph a of this subdivision. Such notice shall be given by the publication of a summary of the application and the date on or about which it will be filed, to be published under regulations to be promulgated by the commission, in such form and in such newspaper or newspapers as will serve substantially to inform the public of such application; and

(ii) persons who have filed a statement with the commission within the past twelve months that they wish to receive all such notices concerning facilities in the area in which the facility is to be located, as primarily proposed or in the alternative locations listed.

3. Inadvertent failure of service on any of the municipalities, persons, agencies, bodies or commissions named in subdivision two shall not be jurisdictional and may be cured pursuant to regulations of the commission designed to afford such persons adequate notice to enable them to participate effectively in the proceeding. In addition, the commission may, after filing, require the applicant to serve notice of the application or copies thereof or both upon such other persons and file proof thereof as the commission may deem appropriate.

4. An application for an amendment of a certificate shall be in such form and contain information as the commission shall prescribe. Notice of such an application shall be given as set forth in subdivision two.

5. If an alternative location, not listed in the application, is proposed in the certification proceeding, notice of such proposed alternative site shall be given as set forth in subdivision two.

6. a. Each application shall be accompanied by a fee of twenty-five thousand dollars to be used to establish a fund (hereafter in this section referred to as the "fund") to defray expenses incurred by municipal parties to the proceeding (except a municipality which is the applicant) for expert witness and consultant fees. The commission shall provide transcripts, reproduce and serve documents, and publish required notices, for municipal parties. Any monies remaining in the fund, after the board has issued its decision on an application under this article and the time for applying for a rehearing and judicial review has expired, shall be returned to the applicant.

b. The twenty-five thousand dollar fee required by subdivision a shall be deposited in one or more separate accounts in one or more banks of the commission's choosing insured by the federal deposit insurance corporation. Notwithstanding any other provision of law to the contrary, the commission shall provide by rules and regulations for the management of the fund, for disbursements from the fund, and for the proper auditing of monies in the fund, which rules and regulations shall be consistent with the purpose of this section to make available to municipal parties monies from such fund for uses specified in this section.

§ 143. Hearing on application for a certificate. 1. Upon the receipt of an application complying with section one hundred forty-two, the chairman shall promptly fix a date for the commencement of a public hearing thereon not less than one hundred eighty nor more than two hundred ten days after such receipt. The place of the hearings shall be designated by the presiding examiner, except that hearings of sufficient duration to provide adequate opportunity to hear direct evidence and rebuttal evidence from residents of the area of the applicant's primary proposed location for the major steam electric generating facility shall be held in such area.

2. On an application for an amendment of a certificate proposing a change in the facility likely to result in any material increase in any environmental impact of the facility or a substantial change in the location of all or a portion of such facility, a hearing shall be held in the same manner as a hearing is held on an application for a certificate. The commission shall promulgate rules, regulations, and standards under which it shall determine whether hearings are required under this subdivision and shall make such determination.

§ 144. Parties to a certification proceeding. 1. The parties to the certification proceedings shall include:

(a) the applicant;

(b) the department of environmental conservation, which shall in any such proceeding present expert testimony and information concerning the potential impact of the proposed facility and any alternative facility or energy source on the environment, and whether and how such facilities would comply with applicable state and municipal environmental protection laws, standards, policies, rules and regulations;

(c) the department of commerce;

(d) the department of health;

(e) the office of planning services;

(f) where the facility or any portion thereof or of any alternate is to be located within its jurisdiction, the Hudson river valley commission;

(h) a municipality entitled to receive a copy of the application under paragraph (a) of subdivision two of section one hundred forty-two, if it has filed with the commission a notice of intent to be a party, within ninety days after the date given in the published notice as the date for filing of the application;

(i) any individual resident in a municipality entitled to receive a copy of the application under paragraph (a) of subdivision two of section one hundred forty-two, if he has filed with the commission a notice of intent to be a party, within ninety days after the date given in the published notice as the date for filing of the application;

(j) any non-profit corporation or association, formed in whole or in part to promote conservation or natural beauty, to protect the environment, personal health or other biological values, to preserve historical sites, to promote consumer interests, to represent commercial and industrial groups or to promote the orderly development of any area in which the facility may be located, if it has filed with the commission a notice of intent to become a party, within ninety days after the date given in the published notice as the date for filing of the application;

(k) any other municipality or resident of such municipality located within a five mile radius of such proposed facility, if it or he has filed with the commission a notice of intent to become a party, within ninety days after the date given in the published notice as the date for filing of the application;

(l) any other municipality or resident of such municipality which the commission or board in its discretion finds to have an interest in the proceeding because of the potential environmental effects on such municipality or person, if the municipality or person has filed with the commission a notice of intent to become a party, within ninety days after the date given in the published notice as the date for filing of the application together with an explanation of the potential environmental effects on such municipality or person; and

(m) such other persons or entities as the commission or board may at any time deem appropriate, who may participate in all subsequent stages of the proceeding.

2. The department of public service shall designate members of its staff who may present evidence concerning any relevant and material matter and shall otherwise participate in proceedings under this article.

3. Any person may make a limited appearance in the proceeding by filing a statement in writing at any time prior to the commencement of the hearing. All papers and matters filed by a person making a limited appearance shall become part of the record. No person making a limited appearance shall be a party or shall have the right to present oral testimony or cross-examine witnesses or parties.

4. The board or the commission may for good cause shown, permit a municipality entitled to become a party under subdivision one, but which has failed to file the requisite notice of intent within the time required, to become a party, and to participate in all sub-

sequent stages of the proceeding.

§ 145. Conduct of the hearing. 1. The hearing shall be conducted by the presiding examiner appointed by the department of public service. An associate hearing examiner shall be appointed by the department of environmental conservation prior to the date set for commencement of the public hearing. His primary responsibility during the pendency of the proceeding shall be to attend hearings and otherwise participate as associate examiner in such proceeding. The associate examiner shall attend all hearings as scheduled by the presiding examiner and inquire into and call for testimony concerning relevant and material matters. The conclusions and recommendations of the associate examiner shall be incorporated in the recommended decision of the presiding examiner, unless the associate examiner prefers to submit a separate report of dissenting or concurring conclusions and recommendations. The testimony presented at such hearing may be presented in writing or orally, provided that the commission may make rules designed to exclude repetitive, redundant or irrelevant testimony. Any governmental agency receiving a request for evidence from a hearing examiner shall comply promptly therewith. A record shall be made of the hearing and of all testimony taken and the cross examinations thereon. The rules of evidence applicable to proceedings before a court shall not apply. The presiding examiner may provide for the consolidation of the representation of parties, other than governmental bodies or agencies, having similar interests. In the case of such a consolidation, the right to counsel to its own choosing shall be preserved to each party to the proceeding provided that the consolidated group may be required to be heard through such reasonable number of counsel as the presiding examiner shall determine. Appropriate regulations may be issued by the commission to provide for prehearing discovery procedures by parties to the proceeding and for consolidation of the representation of parties.

2. A copy of the record shall be made available by the commission at all reasonable times for examination by the public.

3. The presiding examiner shall cause proffered testimony to be received on alternate site and source proposals provided notice of the intent to submit such testimony shall be given within such period as the commission shall prescribe by regulation, which period shall be not less than thirty nor more than sixty days after the commencement of the hearing. Nevertheless, in its discretion, the board or the commission may thereafter cause to be considered other potential sites and sources and cause testimony to be accepted thereon.

4. The chairman of the commission may enter into an agreement with any agency or department of the United States having concurrent jurisdiction over all or part of the location, construction, or op-

eration of a major steam electric generating facility subject to this article with respect to providing for a joint hearing of common issues on a combined record, provided that such agreement shall not diminish the rights accorded to any party under this article.

§ 146. The decision. 1. The board shall make the final decision on an application under this article for a certificate or amendment thereof, upon the record made before the presiding examiner, after receiving briefs and exceptions to the recommended decision of such examiner and to the report of the associate examiner, and after hearing such oral argument as the board shall determine. Petitions for rehearing shall also be considered and decided by the board.

2. The board shall render a decision upon the record either to grant or deny the application as filed or to certify the facility at any site considered at the hearings upon such terms, conditions, limitations or modifications of the construction or operation of the facility as the board may deem appropriate. The board shall issue, with its decision, an opinion stating in full its reasons for its decision. The board shall issue an order upon the decision and the opinion embodying the terms and conditions thereof in full. The board may not grant a certificate for the construction or operation of a major steam electric generating facility, either as proposed or as modified by the board, unless it shall find and determine:

(a) the public need for the facility and the basis thereof;

(b) the nature of the probable environmental impact, including a specification of the predictable adverse effect on the normal environment and ecology, public health and safety, aesthetics, scenic, historic and recreational value, forest and parks, air and water quality, fish and other marine life, and wildlife;

(c) that the facility (i) represents the minimum adverse environmental impact, considering the state of available technology, the nature and economics of the various alternatives, the interests of the state with respect to aesthetics, preservaion of historic sites, forest and parks, fish and wildlife, and other pertinent considerations, (ii) is compatible with the public health and safety; and (iii) will not discharge any effluent that will be in contravention of the standards adopted by the department of environmental conservation or, in case no classification has been made of the receiving waters associated with the facility, will not discharge any effluent that will be unduly injurious to the propagation and protection of fish and wildlife, the industrial development of the state, and public health and public enjoyment of the receiving waters.

(d) that the facility is designed to operate in compliance with applicable state and local laws and regulations issued thereunder concerning, among other matters, the environment, public health and safety, all of which shall be binding upon the applicant, except that the board may refuse to apply any local ordinance, law, resolution or other action or any regulation issued thereunder or any local standard or requirement which would be otherwise applicable if it finds that as applied to the proposed facility such is unreasonably restrictive in view of the existing technology or the needs of or costs to consumers whether located inside or outside of such municipality. The board shall provide the municipality an opportunity to present evidence in support of such ordinance, law, resolution, regulation, or other local action issued thereunder. For the purposes of this article an agreement between the applicant and a municipality in which the proposed facility is to be located, entered into on or before May first, nineteen hundred seventy-one, relating to the location of facilities within the municipality shall be deemed to be and have the force and effect of a local law;

(e) that the facility is consistent with long-rang planning objectives for electric power supply in the state, including an economic and reliable electric system, and for protection of the environment.

(f) that the facility will serve the public interest, convenience, and necessity, provided, however, that a determination of necessity for a facility made by the power authority of the state of New York pursuant to section ten hundred five of the public authorities law shall be conclusive on the board; and

(g) that the facility is in the public interest, considering the environmental impact of the facility, the total cost to society as a whole, the possible alternative sites or alternative available methods of power generation, or alternative available sources of energy as the case may be, both within the state and elsewhere, and the immediacy and totality of the needs of the people of the state for the facility within the context of the need for public utility services and for protection of the environment.

3. A copy of the decision and opinion shall be served on each party personally or by mail.

§ 147. Opinion to be issued with decision. In rendering a decision on an application for a certificate, the board shall issue an opinion stating its reasons for the action taken. If the board has found that any local ordinance, law, resolution, regulation, or other action issued thereunder or any other local standard or requirement which would be otherwise applicable is unreasonably restrictive pursuant to paragraph d of subdivision two of section one hundred forty-six, it shall state in its opinion the reasons therefor.

§ 148. Rehearing and judicial review. 1. Any party aggrieved by any decision on an application for a certificate may apply to the board for a rehearing in the manner provided in section twenty-two of this chapter within thirty days after issuance of the aggrieving decision and thereafter obtain judicial review of such decision in a proceeding as provided in this section. Such proceeding shall be brought in the appellate division of the supreme court of the state in the judicial department embracing the county wherein the facility is to be located, or if the application is denied, the county wherein the applicant has proposed to locate the facility. If such facility is proposed to be located in more than one judicial department, such proceeding may be brought in any one but only one of such departments. Such proceeding shall be initiated by the filing of a petition in such court within thirty days after the issuance of a final decision by the board upon the application for rehearing, together with proof of service of a demand on the commission to file with said court a copy of a written transcript of the record of the proceeding and a copy of the board's decision and opinion. The commission's copy of said transcript, decision and opinion, shall be available at all reasonable times to all parties for examination without cost. Upon receipt of such petition and demand the commission shall forthwith deliver to the court a copy of the record and a copy of the board's decision and opinion. Thereupon the court shall have jurisdiction of the proceeding and shall have power to grant such relief as it deems just and proper, and to make and enter an order enforcing, modifying and enforcing as so modified, remanding for further specific evidence or findings or setting aside in whole or in part such decision. If petitions are filed in more than one court, the court in which a petition was first filed shall retain exclusive jurisdiction of the proceeding, and all other petitions shall be transferred forthwith to said court. Upon motion by any party to the proceeding, or on its own motion, said court may transfer the proceedings to the appellate division in any other judicial department for good cause. The appeal shall be heard on the record without requirement of reproduction. No objection that has not been urged by the party in his application for rehearing before the board shall be considered by the court, unless the failure or neglect to urge such objection shall be excused because of extraordinary circumstances. The findings of fact on which such decision is based shall be conclusive if supported by substantial evidence on the record considered as a whole and matters of judicial notice set forth in the opinion. The jurisdiction of the appellate division of the supreme court shall be exclusive and its judgment and order shall be final, subject to review by the court of appeals in the same manner and form and with the same effect as provided for appeals in a special proceeding. All such proceedings shall be heard and determined by the appellate division of the supreme court and by the court of appeals as expeditiously as possible and with lawful precedence over other matters.

2. The grounds for and the scope of review of the court shall be limited to whether the decision and opinion of the board are:

(a) in conformity with the constitution and the laws of the state and the United States;

(b) supported by substantial evidence in the record and matters of judicial notice properly considered in the opinion;

(c) within the board's statutory jurisdiction or authority;

(d) made in accordance with procedures set forth in this article or established by rule or regulation pursuant to this article.

(e) arbitrary, capricious or an abuse of discretion.

3. Except as herein provided article seventy-eight of the civil practice laws and rules shall apply to appeals taken hereunder.

4. For purposes of section twelve of this chapter, any proceeding or action involving the board shall be deemed to be a proceeding or action involving the commission.

§ 149. Jurisdiction of courts. Except as expressly set forth in section one hundred forty-eight and except for review by the court of appeals of a decision of the appellate division of the supreme court as provided for therein, no court of this state shall have jurisdiction to hear or determine any matter, case or controversy concerning any matter which was or could have been determined in a proceeding under this article or to stop or delay the construction or operation of a major steam electric generating facility except to enforce compliance with this article or the terms and conditions of a certificate issued hereunder.

§ 149-a. Powers of municipalities and state agencies. 1. Notwithstanding any other provision of law, no state agency, municipality or any agency thereof may, except as expressly authorized by this article or the board, require an approval, consent, permit, certificate or other condition for the construction or operation of a major steam electric generating facility with respect to which an application for a certificate hereunder has been filed, other than those provided by otherwise applicable state law for the protection of employees engaged in the construction and operation of such facility, and provided that in the case of a municipality or an agency thereof, such municipality has received notice of the filing of the application therefor.

2. Neither the Hudson river valley commission nor the St. Lawrence-eastern Ontario commission shall hold public hearings for a major steam electric generating facility with respect to which an application hereunder has been filed, provided that such commission has received notice of the filing of such application.

§ 149-b. Long-range electric system planning. Each electric corporation shall prepare and submit annually to the department, at a public hearing upon such notice and at such time and place as the department shall determine, its long-range plan for future operations drawn pursuant to regulations issued by the commission. Such plans shall include:

1. a forecast of demand for the next ten years specifying anticipated load duration, including peak loads;

2. identification of generating capacity to be utilized in meeting such demands, including capacity to be provided by others on a contractual basis;

3. an inventory of (a) all major utility facilities operated by such corporation including the dates for completion and operation of facilities under construction and the dates of the retirement of facilities in operation, and (b) of land owned by the corporation and held for future use as a major steam electric generating facility site;

4. anticipated expenditures for research in the areas of electric generation and transmission and pollution abatement and control during the next year;

5. such additional information as the commission may by regulation require to carry out the purposes of this section.

§ 3. Section one thousand fourteen of the public authorities law, as last amended by chapter two hundred seventy-two of the laws of nineteen hundred seventy, is hereby amended to read as follows:

§ 1014. Public service law not applicable to authority; inconsistent provisions in other acts superseded. The rates, services and practices relating to the generation, transmission, distribution and sale by the authority, of power to be generated from the projects authorized by this title shall not be subject to the provisions of the public service law nor to regulation by, nor the jurisdiction of the department of public service. Except to the extent [article] articles seven and eight of the public service law [applies] apply to the siting and operation of a major utility transmission or major steam electric generating facility as defined therein, the provisions of the public service law and of the conservation law and every other law relating to the department of public service or the public service commission or to the [conservation] department of environmental conservation or to the functions, powers or duties assigned to the division of water power and control by chapter six hundred nineteen, of the laws of nineteen hundred twenty-six, shall so far as is necessary to make this title effective in accordance with its terms and purposes be deemed to be superseded, and wherever any provision of law shall be found in conflict with the provisions of this title or inconsistent with the purposes thereof, it shall be deemed to be superseded, modified or repealed as the case may require.

§ 4. Section eighteen hundred seventy of such law, as amended by chapter two hundred seventy-two of the laws of nineteen hundred seventy, is hereby amended to read as follows:

§ 1870. Public service law not applicable to authority. Except to the extent [article] articles seven and eight of the public service law [applies] apply to the siting and operation of a major utility transmission or major steam electric generating facility, the authority shall not be subject to the provisions of the public service law or to regulation by or the jurisdiction of the department of public service or the public service commission by reason of any contract, agreement or agreement entered by the authority with any power company, any water distribution company or agency or the power authority of the state of New York, or more than one of the above, or by reason of any action taken thereunder by the authority.

§ 5. Paragraph b of subdivision three of section four of the condemnation law, as added by chapter two hundred seventy-two of the laws of nineteen hundred seventy, is hereby amended to read as follows:

b. if the property is to be used for the construction of a major utility transmission facility as defined in section one hundred twenty or major steam electric generating facility as defined in section one hundred forty of the public service law with respect to which a certificate of environmental compatibility and public need has been issued under such law, a statement that such certificate relating to such property has been issued and is in force.

§ 6. Section twelve hundred thirty of the public health law is hereby amended, by adding thereto a new subdivision, to be subdivision eight, to read as follows:

8. In the case of a major steam electric generating facility, as defined in section one hundred forty of the public service law, for the construction or operation of which a certificate is required under article eight of such law, an applicant shall apply for and obtain such certificate in lieu of filing plans and reports and obtaining a permit under this section. Any reference in this article to a permit under this section shall, in the case of such major steam electric generating facility, be deemed for all purposes to refer to such certificate, provided that nothing herein shall limit the authority of the departments of health and environmental conservation to monitor the environmental and health impacts resulting from the operation of such major steam electric generating facility and to enforce applicable provisions of the public health and environmental conservations laws and the terms and conditions of the certificate governing the environmental and health impacts resulting from such operation.

§ 7. Paragraph (j) of subdivision two of section twelve hundred seventy-seven of such law, as added by chapter nine hundred two of the laws of nineteen hundred sixty-six, is hereby amended to read as follows:

(j) Consider for approval plans or specifications for air cleaning installations or any part thereof submitted to him pursuant to the rules of the board, and inspect the installation for compliance with the plans or specifications; provided that in the case of a major steam electric generating facility, as defined in section one hundred forty of the public service law, for which a certificate is required pursuant to article eight of such law, such approval functions shall be performed by the board and such inspection functions shall be performed by the public service commission; provided further that nothing herein shall limit the authority of the departments of health and environmental conservation to monitor the environmental and health impacts resulting from the operation of such major steam electric generating facility and to enforce applicable provisions of the public health and environmental conservation laws and the terms and conditions of the certificate governing the environmental and health impacts resulting from such operation.

§ 8. This act shall take effect July first, nineteen hundred seventy-two and shall continue in full force and effect only until January first, nineteen hundred seventy-nine, provided, however, that the provisions of this act shall remain operative and continue in full force and effect with regard to applications filed on or before December thirty-first, nineteen hundred seventy-eight.

# APPENDIX

## MEASURES FOR ENERGY CONSERVATION

The material below was taken from *The Potential for Energy Conservation*, a governmental interagency staff study undertaken by the Interagency Working Group under the leadership of the Director of the Office of Emergency Preparedness and made available in October, 1972.

The following charts present a list of energy conservation measures for the short-term (1972-1975), mid-term (1976-1980) and the long-term (beyond 1980) for the Transportation, Residential/Commercial, Industrial and Electric Utility sectors. The charts also indicate estimated maximum attainable energy savings, possible means for implementing each conservation measure, and pros and cons.

Savings figures refer to annual savings as of the last year of the time period indicated, i.e., 1975 for the short-term, 1980 for the mid-term and 1990 for the long-term. The savings associated with the Electric Utility sector, however, have already been assumed in the projections of energy consumption. Percentages of demand refer to savings in projected sector consumption. Electrical energy savings are included for each of the sectors.

One means of implementing many of the suggested conservation measures might be through a significant increase in fuel prices either by increasing energy use taxes or by regulation. Although there is considerable uncertainty about the price elasticity of energy as a whole, the best informed judgment is that energy demand is quite responsive to price changes in the mid-term and long-term and not very responsive in the short-term. A price increase is not repeated throughout the charts, although it may be understood as one of the possible means of implementing many of the suggested conservation measures.

Transportation
Short Term (1972-1975)

| Measure | Est. Annual Savings Trillion BTU by 1975 and Percentage of Sector Demand | Implemented by | Pro | Con |
|---|---|---|---|---|
| General: | | | | |
| --Energy conservation - public awareness. | | Government sponsored programs | Inexpensive. Critical to development of needed popular support for stronger measures. | |
| --Inject energy conservation issue into appropriate national programs (e.g., environment, health, safety, urban reform, etc. | | Cooperative programs. | | |
| --Stimulate participation of citizens' groups in transportation planning and development. | | Cooperative programs. | | |
| --Enlist industry cooperation in promoting public energy conservation attitudes. | | Persuasion. | | |
| Increase Energy Efficiency: | | | | |
| --Develop energy-efficiency standards. | | Government program. | | |
| --Air - increase loading factors. | 125 | Persuasion/regulation. | Highly effective. No cost to Federal Government. | Airlines profits and services may decline. |
| --Trucks - ban urban operation during high congestion periods. | 75 | Persuasion/regulation. | Inexpensive to implement. | Resistance from merchants and trucking companies. |
| ..Loading regulations. | | Regulation. | | |
| ..Freight consolidation. | | Subsidy/regulation. | | |

Transportation
Short Term (1972-1975)

| Measure | Est. Annual Savings Trillion BTU by 1975 and Percentage of Sector Demand | Implemented by | Pro | Con |
|---|---|---|---|---|
| --Auto - promote development and sales of smaller engines/vehicles. Improved propulsion and drive train system and tires. | 900 | Persuasion/tax incentives. | Highly effective in relation to cost. Also serves to decrease pollution. | Several of these pressures require the availability of suitable mass transit. |
| ..Increased car pooling. | | Tax/toll. | | |
| ..Improve driver procedure. | | Cooperative program. | | |
| ..Improved maintenance. | | Cooperative/regulatory. | | |
| ..Decreased operation in in congested areas. | | Increase parking tax, peak hour road use toll/tax, 4-day week/staggered work hours (balanced against pooling) regulations, free fringe parking. | | |
| --Other measures to improve traffic flow. | | | Effective measures that should not meet much resistance. | Cost of monitoring and control systems is high (est. $50,000 per mile). |
| ..Reverse lane operation. | | Regulation. | | |
| ..Driver advisory displays. | | Subsidy. | | |
| ..Helicopter reports. | | Subsidy. | | |
| ..Ramp control. | | Subsidy. | These measures may be critical to improving motorized mass transit. | |
| ..Traffic monitoring systems. | | Subsidy. | | |
| ..Pedestrian control/walkways. | | Subsidy/regulation. | | |
| ..Reversible one-way streets. | | Regulation. | | |
| ..Embargoed traffic zones. | | Regulation. | | |
| <u>Improved Balance Between Modes:</u> | | | | |
| --Urban passenger. | 400 | | | |

Transportation
Short Term (1972-1975)

| Measure | Est. Annual Savings Trillion BTU by 1975 and Percentage of Sector Demand | Implemented by | Pro | Con |
|---|---|---|---|---|
| .. Decrease mass transit fares. | | Subsidy. | Particularly important in increasing mobility of the poor. Major effect on fuel consumption, convertibility, and pollution. | High cost. Possible detrimental effect on the automobile and related industries. |
| .. Improve motorized mass transit service. | | Subsidy (matching grants). | | |
| .. Demonstration mass transit projects. | | Federal programs. | | |
| .. Increased fringe parking. | | Regulation. | | |
| .. Car pooling to trunk lines. | | Cooperative. | | |
| .. Priority bus lanes/zones. | | | | |
| .. Bus priority at intersections. | | | | |
| --Intercity passenger. | 175 | | Increased use of rail is opposed to air has tremendous effect on fuel consumption. Availability of sufficient rail service may prove critical in times of inadequate petroleum supply. | Detrimental effect on air industries. |
| .. Ban subsidy of short flights by long flights. | | Regulation. | | |
| .. Improve rail service. | | Subsidy. | | |
| .. Demonstration rapid surface systems. | | Matching grants. | | |
| .. Realign regulation and R&D policy to insure desired modal balance. | | Policy/regulation. | | |
| --Intercity freight. | 175 | Subsidy. | | Possible strong resistance from highway and travel lobbies. |
| .. Improve rail service. | | Subsidy/tax incentives. | | |
| .. Encourage piggy backing. | | | | |
| .. Freight consolidation and containerization. | | | | |

Transportation
Short Term (1972-1975)

| Measure | Est. Annual Savings Trillion BTU by 1975 and Percentage of Sector Demand | Implemented by | Pro | Con |
|---|---|---|---|---|
| Decrease Demand for Transportation: | | | | |
| --Increased incentives to walking (e. g., attractive walkways, bicycle paths, sheltered rest and refreshment). | 50 | Subsidy. | Very low cost in proportion to benefits. | Possible right of way problem and resistance from highway groups. |
| --Promote use of nearby facilities (e. g., dining, stores, recreation). | | Persuasion. | Inexpensive. Should prove valuable in furthering needed urban redesign. | May hurt tourist trade. |
| --Promote use of communications facilities in lieu of travel. | | Persuasion. | | |
| TOTAL | 1900 (10%) | | | |

Residential/Commercial
Short Term (1972-1975)

| Measure | Est. Annual Savings Trillion BTU by 1975 and Percentage of Sector Demand | Implemented by | Pro | Con |
|---|---|---|---|---|
| Reduce heat loss in winter and heat gain in summer in existing residential dwellings through addition of insulation storm windows, caulking, humidifiers, attic fans, etc. | | Income tax law revisions to make expenditures for energy conservation deductible. | Provide inducement to homeowners at small cost to the Federal Government. | Regressive tax. |
| | | Educational programs. | Moderate cost. | |
| | | Government insured loans for energy conservation measures. | Strong incentive demonstrating serious intent of government. | Costly to administer. |
| Encourage purchase of more efficient appliances. | | Require that a nameplate giving energy consumption be affixed to all new appliances. Require same information on price tags and advertisements. | Permits determination of energy costs by buyer. | Some policing necessary to verify that data shown is correct. |
| | | | Encourage manufacturer to improve the efficiency of their appliances. | Impose additional burden on manufacturer and for retailer. |
| | | Provide financial support to non-government institutions to test and publicize energy consumption of appliances. | Provides unbiased information to consumer on energy costs. | Requires government funding. |
| Encourage adoption of good energy conservation practices in the "operation" of the home. | | Educational programs. | Low cost program which reduces energy bills and consumption. | Amount of energy to be saved is modest. |
| TOTAL | 200 (1%) | | | |

Industry
Short Term (1972-1975)

| Measure | Est. Annual Savings Trillion BTU by 1975 and Percentage of Sector Demand | Implemented by | Pro | Con |
|---|---|---|---|---|
| Economic incentive to upgrade processes and replace inefficient equipment. | 1600 - 3200<br>5 - 10% | Energy price rise by energy use tax. | Cut energy demand using existing technology. | Requires tax or tax incentive, increases production costs which may cause increased imports and decreased exports. |
| Encourage research in efficient technologies. | No effect in short term. | Government research funding or tax incentive. | May accelerate existing research. | Requires increased government spending. |
| Encourage recycling and reusing of component materials. | 300<br>1.0% | Tax incentive for use of recycled material, system of standards and regulations to orient original equipment design toward recycling, research in recycling technology. | Conserves resources, reduces dependence on foreign suppliers (improves balance of payments, national security). | Energy savings is slight. |

Electric Utilities
Short Term (1972-1975)

| Measure | Est. Annual Savings Trillion BTU by 1975 and Percentage of Sector Demand | Implemented by | Pro | Con |
|---|---|---|---|---|
| Smooth out daily demand cycle by shifting heavy loads to off peak hours. | 500<br>2% | Restructure rates for heavy uses, exclude some heavy uses from peak hour use by regulation. | Reduce use of inefficient peaking equipment, reduce capital requirements in peaking generators. | Shift in industrial load to off peak hours requires people to work odd hours. |
| Decrease electricity demand. | See other sections. | See other sections for recommendations. | May increase efficiency of fuel use. | If shifted demand to other fuels, may lose centralized pollution control. |
| Facilitate new construction, and alleviate construction delays. | 500<br>2% | Intensify government efforts to inform and work with labor environmental and conservation groups. | Accelerate replacement of old inefficient equipment. | |
| | | Develop regional, industry-wide clearinghouses with possible regulation commission participation to spread out new plant and equipment orders over a reasonable time period. | Reduce periodic strain on construction labor markets and equipment manufacture, reducing incidence of strikes and new plant and equipment breakdown. | Reduces planning flexibility of individual utilities. |
| | | Make equipment manufacturers fully liable for equipment failures. | Provide incentive for quality control. | |
| | | Remove equipment from rate base if excessively unavailable. | Provide incentive for high quality maintenance and supervision of construction. | |

Electric Utilities
Short Term (1972-1975)

| Measure | Est. Annual Savings Trillion BTU by 1975 and Percentage of Sector Demand | Implemented by | Pro | Con |
|---|---|---|---|---|
| | | Revise siting review procedures to include participation of interested parties, including Federal, State, and local environmental agencies, and conservation groups. | Obtain agreement on sites before construction begins. | |
| | | Intensify government efforts to resolve controversies concerning safety of nuclear plants and waste storage facilities. | Remove public opposition to nuclear plant construction and operation. | |

Transportation
Mid Term (1976-1980)

| Measure | Est. Annual Savings Trillion BTU by 1980 and Percentage of Sector Demand | Implemented by | Pro | Con |
|---|---|---|---|---|
| General | | | | |
| --Increase Fuel Costs | Reflected in estimates cited below. | Tax | Important source of revenue for transportation programs, uncomplicated. | May not discourage extravagant use of transportation. Disadvantage to the poor unless inexpensive. Adequate mass transit is available. |
| Increase Energy Efficiency | | | | |
| --Enforce Efficiency Standards | | Regulation | Readily enforced. | |
| --Auto | | | | |
| . Selective registration tax on size, power, power attachments. | Energy saving offset by emission control fuel penalty. | Tax | | Possible loss in sales to U. S. auto industry and consequent balance of payments deficit increases. |
| . Select and encourage pilot implementation of the most promising alternatives to internal combustion engine. | | Subsidy/Tax Incentives | May serve to strengthen U. S. auto industry in long run. | |
| . Controls on auto size/loading and entry into central business districts. | 2,400 | Regulation | Higher effective. | Requires suitable mass transit capacity. |
| . Improved propulsion systems, increased use of loss tires. | | | | |
| --Air | | | | |
| . Require operating procedures designed to minimize energy consumption. | 100 | Regulation | Enormous increases in efficiency are possible. | Involves tradeoffs in speed, convenience, etc. vs. energy demand. Possible negative effect on air industry. |

Transportation
Mid Term (1976-1980)

| Measure | Est. Annual Savings Trillion BTU by 1980 and Percentage of Sector Demand | Implemented by | Pro | Con |
|---|---|---|---|---|
| --Buses | | | | |
| ..Improved propulsion systems | | | | |
| ..Increase measures designed to improve traffic flow. | | | | |
| --Trucks | | | | |
| ..Improved engines (possibly retrofitting fleets). | 300 | Subsidy/Regulation | | |
| ..Controls on deliveries. | | | | |
| ..Increased consolidation. | | | | |
| Improved Balance Between Modes | | | | |
| --Urban Passenger | 300 | | | |
| ..Improved feeder service "Dial-A-Bus." | | Subsidy/Matching Grants | Potentially large savings in energy and large benefits in increased environmental quality and mobility. | High cost. Requires careful specification of community objectives and needs and close cooperation between many diverse groups. |
| ..Increased bans on autos in center city. | | Regulation | | |
| ..Improved arterial mass transit. Motorized surface and rapid transit. | | Subsidy/Matching Grants | | |
| ..Demonstration mass transit systems. | | Matching Grants | | |
| ..Increased conversion of center city areas into pedestrian-oriented clusters. | | Matching Grants | | |
| --Intercity Passenger | 900 | | | |
| ..Improved rail networks. | | Subsidy | Very large potential savings at minimal cost. | May require increased government involvement. |
| ..Multi-modal development, implementation, regulation. | | | | |

Transportation
Mid Term (1976-1980)

| Measure | Est. Annual Savings Trillion BTU by 1980 and Percentage of Sector Demand | Implemented by | Pro | Con |
|---|---|---|---|---|
| --Intercity Freight<br>..Improved rail freight handling procedures and service/containerization. | 600 | Subsidy | Large payoff for small cost. | |
| <u>Decreased Transportation Demand</u> | | | | |
| --Demonstration and Urban Refurbishing Projects | 200 | Matching Grants | Urgent problem that is intimately linked to transportation. | Expensive. |
| --Long Range Urban/Suburban Planning<br>..Building restrictions.<br>..Continued development of aids to walking. | | Matching Grants | | |
| TOTAL SAVINGS | 4,800 (20%) | | | |

Residential/Commercial
Mid Term (1976-1980)

| Measure | Est. Annual Savings Trillion BTU by 1980 and Percentage of Sector Demand | Implemented by | Pro | Con |
|---|---|---|---|---|
| --Continuation and extension of measures listed for short term. | 1,100 (3%) | | | |
| --Design and build new homes so that heat loss in winter and heat gain in summer will be a minimum. | | | | |
| . . Increased and improved insulation, caulking, double glazed windows, and reduced window area. | | More stringent FHA minimum property standards. | Builders will follow FHA standards. | Raises cost of homes. Construction industry might resist. |
| | | Federal or state law requiring all new homes meet FHA minimum property standards. | | (Same as above.) |
| | | Require an energy assessment on all government connected buildings by architect before construction is undertaken. | Will encourage adoption of designs with low energy requirements. Serves as model for nation. | Increases cost of design work. |
| | | Require inspection and energy appraisal by independent engineer to be furnished to buyer before sale. | (Same as above.) | |
| | | Government connected buildings to be designed for minimum life cycle costs. | Serve as model for nation. | Increase construction costs. |

Residential/Commercial
Mid Term (1976-1980)

| Measure | Est. Annual Savings Trillion BTU by 1980 and Percentage of Sector Demand | Implemented by | Pro | Con |
|---|---|---|---|---|
| | | Require plaque in new homes receiving FHA mortgage guarantee stating insulation features and energy requirement for heating, cooling, and major appliances. | House buyer can estimate heating and cooling costs. | Plaque installation and data collection costs will be passed on to purchaser. |
| | | Require independent inspection by a licensed engineer. Report on insulation and energy requirements of new houses. | (Same as above.) | Increased costs to home buyer. |
| --Improve efficiency of furnaces and air conditioning units. | | Legislation setting minimum efficiency levels. | If done in cooperation with trade associations likelihood of major dislocations within industry is small. | Increase costs. Manufacturers have no incentive to exceed minimum required level. |
| --Reduce energy consumption of all appliances. | | Install in the kitchen of all homes an indicating watt-meter showing instantaneous energy consumption of the household. | Will call attention to abnormally high electricity usage so that corrective action may be taken. | Costs about $100 per house installed. |
| | | Government procurement policy to reflect the cost of energy over the life of the appliance. | Encourage manufacturer to improve the efficiency of their appliances. | Will increase purchase price of appliances. |
| | | Provide financial support to non-government institutions to test and publicize energy consumption of appliances. | Provides unbiased information to consumer on energy costs. | Requires government funding. |

Residential/Commercial
Mid Term (1976-1980)

| Measure | Est. Annual Savings Trillion BTU by 1980 and Percentage of Sector Demand | Implemented by | Pro | Con |
|---|---|---|---|---|
| --Provide single large efficient air conditioning, heating plant to serve a number of houses or commercial building. | | Tax incentive to housing developers. | Reduces energy costs and consumption. | May involve higher capital and maintenance costs due to need to install distribution system (supply and return) for hot and chilled water. |
| --Reduce heat added in air handling system by minimizing reheat for humidity control, intake of ourside air and duct velocities. | | Require efficient designs in all buildings in which the Federal Government has an interest. | Sets example for the building industry. | To reduce intake of outside air may require revisions to local codes. |
| | | Federal procurement specifications requiring high efficiency designs. | May induce manufacturers to offer the public high-efficiency designs. | Increases acquisition costs to government. |
| --Recovery and utilization of heat rejected by furnaces. water heaters, and condensers of air conditioning units and in air exhausted from building. | | Government participation in research and development programs, followed by revised FHA standards. | Manufacturers will meet improved FHA standards. | Savings in energy costs may not offset capital outlay except over the long term. |
| | | R&D program to develop better methods storing thermal energy. | | |
| --Wider use of district heating for urban multi-family dwelling, particularly where steam is obtained from exhaust of turbine generators. | | Provide incentives to municipalities to require all buildings located near district heating mains to utilize same. | Substantial reduction in number of apartment house boilers. Reduction in amount of fuel consumed in city. | Requires conversion of some existing heating systems. |
| --Wider use of total energy concept for multi-family dwellings. | | Tax and other incentives for selection of total energy concept. | Induces owner to select system offering higher overall efficiencies and lower total cost for energy. | Current systems use gas and gas is in short supply. |

Residential/Commercial
Mid Term (1976-1980)

| Measure | Est. Annual Savings Trillion BTU by 1980 and Percentage of Sector Demand | Implemented by | Pro | Con |
|---|---|---|---|---|
| --Discourage building of single-family dwellings. | | Remove tax advantage for purchasers of new single detached dwelling. | Reduces construction of single-family dwellings which are less energy efficient than multi-dwelling units. | Runs counter to the long established goal of most Americans to own their own house, discourages replacing old houses with new houses. |
| Total space heating and cooling savings | 1,600 (4%) | | | |
| --Reduce energy required for hot water heaters. | | | | |
| ..Offer more efficient designs (no continuous gas pilot, thicker shell insulation, etc.) | | Legislation setting minimum efficiency performance levels. | If done in cooperation with trade association, there will be little dislocation within industry. | Effective policing may be costly. |
| ..Recover heat in stack gases and in hot water going down drains. | | Building codes requiring use of heat recovery systems. | Reduces energy requirements for heating. | Increased capital costs. |
| ..Provide devices to keep heat exchange surfaces clean. | | Legislation or code requiring cleaning device on all furnaces. | Will increase efficiency by 50 pecent. (NBS estimate) | Increased capital costs. |
| ..Limit hot water consumption of washing machines and dishwashers. | | Set standards for new equipment under some level of heat usage. | | |
| | | Impose taxes on hot water detergents so as to encourage use of cold water detergents. | Discourage manufacturer of inefficient appliances. | No incentive for manufacturer to improve performance beyond the requirements of standards. |
| Total hot water heating savings | 250 (1%) | | | |

Residential/Commercial
Mid Term (1976-1980)

| Measure | Est. Annual Savings Trillion BTU by 1980 and Percentage of Sector Demand | Implemented by | Pro | Con |
|---|---|---|---|---|
| --Reduce energy required for cooking. | | | | |
| .. Redesign range burners so that more heat enters pot and less is lost to the ambient air. | | Federal procurement specification on gas and electric ranges and ovens. | Sets example for industry. Some manufacturers may add high efficiency range to their line. | Compliance is not mandatory.<br>Increases cost of range and ovens bought by government. |
| .. Require thicker insulation in oven walls. | | Legislation prohibiting sale of ranges with inefficient burners and ovens with inadequate insulation. | More effective than measures limited to government procurement. | Increases cost of range and oven.<br>Requires special pot size and bottom shapes. |
| Total cooking savings | 50 | | | |
| --Reduce energy consumption of refrigerators | 100 | Government participation in R&D program. | | Raises price of typical refrigerator by about $30. |

Residential/Commercial
Mid Term (1976-1980)

| Measure | Est. Annual Savings Trillion BTU by 1980 and Percentage of Sector Demand | Implemented by | Pro | Con |
|---|---|---|---|---|
| --Reduce energy required for lighting and miscellaneous appliances. | 500 (1%) | Advance the official time by one hour in winter and two hours in summer. | Reduces requirement for artificial lighting. No administrative cost. | Inconvenient for some groups, e.g., farmers. |
| | | R&D program to develop durable switch which will automatically shut off the lights in a room after the last occupant has left. | Reduces unnecessary energy consumption. | More expensive than ordinary switches. |
| | | Prohibit installation in new FHA homes of incandescent type lamps for kitchens, bathrooms and yard lighting. | Provide more energy efficient lighting. | Increases lighting fixtures costs. Reduces consumer choice. |
| | | R&D program to develop a new fluorescent lamp that is interchangeable with existing incandescent lamps. | Consumes only 1/3 electricity of incandescent lamps. | Higher initial price. |
| | | Revise standards so as to reduce lighting levels. | Reduces air conditioning requirements also. | |
| | | Establish codes for minimum window and skylight areas (double or triple pane) in offices and factories to provide more natural light. | Reduces requirements for artificial lighting. | Heat flow through large glass windows may exceed energy requirements of lamps. |
| Comparable measures for commercial establishments | 1,500 (4%) | | | |
| Total savings | 5,100 (14%) | | | |

Industry
Mid Term (1976-1980)

| Measure | Est. Annual Savings Trillion BTU by 1980 and Percentage of Sector Demand | Implemented by | Pro | Con |
|---|---|---|---|---|
| Economic incentive to upgrade processes and replace inefficient equipment. | 3,800 - 5,600<br>10 - 15% | Energy price rise by energy use tax. | Cut energy demand using existing technology. | Increases production costs which may cause increased imports and decreased exports. |
| Encourage research in efficient technologies. | Some effect. | Government research funding or tax incentive. | May accelerate existing research. | Requires increased government spending. |
| Encourage recycling and reusing of component materials. | 750<br>2% | Tax incentive for use of recycled material, system of standards and regulations to orient original equipment design toward recycling, research in recycling technology. | Conserves resources, reduces dependence on foreign suppliers (improves balance of payments, national security). | Energy savings is slight. |

Electric Utilities
Mid Term (1976-1980)

| Measure | Est. Annual Savings Trillion BTU by and Percentage of Sector Demand | Implemented by | Pro | Con |
|---|---|---|---|---|
| Smooth out daily demand cycle by shifting heavy loads to off peak hours. | 600<br>2% | Restructure rates for heavy uses, exclude some heavy uses from peak hour use by regulation. | Reduce use of inefficient peaking equipment, reduce capital requirements in peaking generators. | Shift in industrial load to off peak hours requires people to work odd hours. |
| Decrease electricity demand. | See other sections. | See other sections for recommendations. | May increase efficiency of fuel use. | If shifted demand to other fuels, may lose centralized pollution control. |
| Facilitate new construction and alleviate construction delays. | 500<br>2% | Itensify government efforts to inform and work with labor, environmental and conservation groups. | Accelerate replacement of old inefficient equipment. | |
| | | Develop regional, industry-wide clearinghouses with possible regulation commission participation to spread out new plant and equipment orders over a reasonable time period. | Reduce periodic strain on construction labor markets and equipment manufacture, reducing incidence of strikes and new plant and equipment breakdown. | Reduces planning flexibility of individual utilities. |
| | | Make equipment manufacturers fully liable for equipment failures. | Provide incentive for quality control. | |
| | | Remove equipment from rate base if excessively unavailable. | Provide incentive for high quality maintenance and supervision of construction. | |

Electric Utilities
Mid Term (1976-1980)

| Measure | Estimated Annual Savings by and Percentage of Sector Demand | Implemented by | Pro | Con |
|---|---|---|---|---|
| Support Electric Research Council research funding and contracting corporation. | Slight effect. | Have Federal Power Commission provide information, guidelines and recommended actions to local public utility commissions. | Will provide utility industry with a means for setting research priorities and funding research. | Funds must be approved by local PUC's. |

Transportation
Long Term (Beyond 1980)

| Measure | Est. Annual Savings Trillion BTU by 1990 and Percentage of Sector Demand | Implemented by | Pro | Con |
|---|---|---|---|---|
| Increased Energy Efficiency: | 1500 | | | |
| --Advanced propulsion systems. | | R&D. | Prospects look good for increasing efficiency by 100 percent. | Implementation data uncertain. |
| --Non-Petroleum Engines. | | R&D. | Free from dependence on petroleum reserves. | Possibly no overall savings in energy. |
| --Advanced Traffic Control. | | R&D. | Prospects good for low cost means of improving traffic flow/safety. | |
| --New Transportation Systems. | | R&D. | | |
| Better Balance Between Modes: | 4000 | | | |
| --Dual mode personal rapid transit. | | R&D. | The position of the U.S. transportation industries in the world market place may well depend on investment in R&D. | Many proposed systems have lower energy efficiencies. |
| --High speed surface transit system. | | | | |
| --New freight handling/distribution systems. | | R&D. | | |
| --New people movers. | | R&D. | | |
| Decrease Demand: | 2500 | | | |
| --Rationing. | | Regulation. | May be the only measure possible. | Highly unpalatable last resort. |
| --Planned urban development/reconstruction (clustering). | | Matching grants. | Enormous potential. | High cost. |
| TOTAL | 8000 (25%) | | | |